T0139885

Springer Theses

Recognizing Outstanding Ph.D. Research

Aims and Scope

The series "Springer Theses" brings together a selection of the very best Ph.D. theses from around the world and across the physical sciences. Nominated and endorsed by two recognized specialists, each published volume has been selected for its scientific excellence and the high impact of its contents for the pertinent field of research. For greater accessibility to non-specialists, the published versions include an extended introduction, as well as a foreword by the student's supervisor explaining the special relevance of the work for the field. As a whole, the series will provide a valuable resource both for newcomers to the research fields described, and for other scientists seeking detailed background information on special questions. Finally, it provides an accredited documentation of the valuable contributions made by today's younger generation of scientists.

Theses are accepted into the series by invited nomination only and must fulfill all of the following criteria

- They must be written in good English.
- The topic should fall within the confines of Chemistry, Physics, Earth Sciences, Engineering and related interdisciplinary fields such as Materials, Nanoscience, Chemical Engineering, Complex Systems and Biophysics.
- The work reported in the thesis must represent a significant scientific advance.
- If the thesis includes previously published material, permission to reproduce this must be gained from the respective copyright holder.
- They must have been examined and passed during the 12 months prior to nomination.
- Each thesis should include a foreword by the supervisor outlining the significance of its content.
- The theses should have a clearly defined structure including an introduction accessible to scientists not expert in that particular field.

More information about this series at http://www.springer.com/series/8790

Jemmyson Romário de Jesus

Proteomic and Ionomic Study for Identification of Biomarkers in Biological Fluid Samples of Patients with Psychiatric Disorders and Healthy Individuals

Doctoral Thesis accepted by
the University of Campinas, Campinas, Brazil

 Springer

Author
Dr. Jemmyson Romário de Jesus
Institute of Chemistry
University of Campinas
Campinas, Brazil

Supervisor
Prof. Marco Aurélio Zezzi Arruda
Institute of Chemistry
University of Campinas
Campinas, Brazil

ISSN 2190-5053   ISSN 2190-5061 (electronic)
Springer Theses
ISBN 978-3-030-29475-5   ISBN 978-3-030-29473-1 (eBook)
https://doi.org/10.1007/978-3-030-29473-1

This Springer imprint is published by the registered company Springer Nature Switzerland AG
The registered company address is: Gewerbestrasse 11, 6330 Cham, Switzerland

Supervisor's Foreword

Proteomic and Ionomic Study for Identification of Biomarkers in Biological Fluid Samples of Patients with Psychiatric Disorders and Healthy Individuals is a compilation of different proposals in terms of omics strategies, within the context of the Springer Thesis, and regarding to Ph.D. Thesis of Dr. Jemmyson Romário de Jesus, with collaboration of Prof. José Luis Capelo-Martinez (Bioscope Group, Caparica, Portugal). This book is organized in three chapters, focusing on studies to identify blood-group biomarkers of patients with bipolar disorder (BD), healthy controls, including non-family members (HCNF) and family members (HCF), patients with schizophrenia (SCZ), and patients with other disorders (OD). Additionally, a new analytical method was developed to study the urine proteome applied in the discovery of biomarkers of human disease through urine analysis. Afterwards, Chap. 1 presents the evaluation of three different methods to simplify the proteome of the blood serum, such as depletion of abundant proteins using chemical agents (acetonitrile, ACN, and dithiothreitol, DTT), depletion of abundant proteins using magnetic nanoparticles (MNPs), and the equalization of proteins using commercial ProteoMiner (PM) kit. In Chap. 2, two methods were tested for the decomposition of the sample: microwave assisted decomposition (conventional method) and fast decomposition assisted by high power ultrasound (proposed method), for evaluating ionomics from SCZ, BD and OD patients compared to controls (HCF and HCNF). Finally, Chap. 3 deals with a new membrane-based methodology to extract proteins present in urine free from interferers. After optimization of some parameters, such as polymer membrane composition, membrane pore size, urine flow rate and medium pH, a significant classification of genders was obtained, according to statistical data. For those professionals involved in omics-science, analytical, biological, toxicological, medical, among others areas, this book will be a useful reference, while providing valuable information from a transdisciplinary point of view.

Campinas, Brazil   Prof. Marco Aurélio Zezzi Arruda
October 2019

Abstract

Bipolar disorder (BD) is a complex psychiatric disorder that affects thousands of people in the world. Some patients with BD are misdiagnosed as unipolar depression and therefore are treated ineffectively. In this sense, an exploratory analysis using comparative serum (proteomic and ionomic) strategies in the blood serum of patients with BD, healthy controls, including non-family (HCNF) and family (HCF), patients with schizophrenia (SCZ), and patients with other disorders (OD) was performed in order to determine biomarkers of BD. For the proteomic study, three methods were evaluated to simplify the proteome of the blood serum: depletion of abundant proteins using chemical agents (acetonitrile, ACN, and dithiothreitol, DTT); depletion of abundant proteins using magnetic nanoparticles (MNPs) and the equalization of proteins using commercial ProteoMiner (PM) kit. For the ionomic study, two methods were tested for the decomposition of the sample: microwave assisted decomposition (conventional method) and fast decomposition assisted by high power ultrasound (proposed method). As a result of the proteomic study, the PM method presented as the best strategy to remove the majority proteins. By comparing 2-D DIGE gel images, 37 protein spots were found to be differentially abundant ($p < 0.05$, Student's t-test), which exhibited a variation of the regulation factor ≥ 2 times the mean value of the intensities of serum spots of SCZ, BD and OD patients compared to controls (HCF and HCNF). From these detected spots, 13 different proteins were identified: ApoA1, ApoE, ApoC3, ApoA4, Samp, SerpinA1, TTR, IgK, Alb, VTN, TR, C4A and C4B. As a result of the ionomic study, the method proposed using ultrasound presented the best conditions for the extraction of Zn, Cu, Fe, Li, Cd and Pb from serum. From the application of the proposed method, the recoveries of analytes ranged from 80 to 121%, with relative standard deviation ranging from 3–10% (n = 3). Thus, the optimum conditions were applied in the serum samples of BD, SCZ, and healthy individuals, where the absence of Pb and Cd was observed for all the evaluated samples, and significant differences in the concentration of Zn, Cu, Li and Fe between evaluated groups. For BD, a high level of metal concentration was observed, while for the SCZ group, all metals were found at low levels. In addition, a new membrane based methodology was developed to extract proteins present in urine free of interferers. In this study, after optimization of some parameters, such as polymer membrane composition, membrane pore size, urine flow rate and medium pH, the better results were observed using nitrocellulose membrane, with a pore size of 0.22 μm, and urine flow rate of 0.25 mL min^{-1}, in addition to using a pH 3 to obtain a retention of membrane urinary protein greater than 90%. In addition, the methodology developed was applied to a gender classification study. As a result, a significant classification was obtained, according to statistical data.

Parts of this Thesis have been Published in the Following Journal Articles

De Jesus, Jemmyson Romário; Guimarães, Ivanilce Cristina; Arruda, Marco Aurélio Zezzi. Quantifying proteins at microgram levels integrating gel electrophoresis and smartphone technology. **Journal of Proteomics**, v.198, p. 45–49, 2019.

De Jesus, Jemmyson Romário; Santos, Hugo Miguel; López-Fernández, Hugo; Lodeiro, Carlos; Arruda, Marco Aurélio Zezzi; Capelo, José Luis. Ultrasonic-based membrane aided sample preparation of urine proteomes. **Talanta**, v. 178, p. 864–869, 2018.

De Jesus, Jemmyson Romário; da Silva Fernandes, Rafael; de Souza Pessôa, Gustavo; Raimundo, Ivo Milton; Arruda, Marco Aurélio Zezzi. Depleting high-abundant and enriching low-abundant proteins in human serum: an evaluation of sample preparation methods using magnetic nanoparticle, chemical depletion and immunoaffinity techniques. **Talanta**, v.170, p.199–209, 2017.

De Jesus, Jemmyson Romário; Galazzi, Rodrigo Moretto; de Lima, Tatiani Brenelli; Banzato; Cláudio Eduardo Muller; de Almeida Lima e Silva, Luiz Fernando; de Rosalmeida Dantas, Clarissa; Gozzo, Fábio Cézar; Arruda, Marco Aurélio Zezzi. Simplifying the human serum proteome for discriminating patients with bipolar disorder of other psychiatry conditions. **Clinical Biochemistry**, v. 50, p.1118–1125, 2017.

De Jesus, Jemmyson Romário; Pessôa, Gustavo de Souza; Sussulini, Alessandra; Martínez, José Luis Capelo; Arruda, Marco Aurélio Zezzi. Proteomics strategies for bipolar disorder evaluation: from sample preparation to validation. **Journal of Proteomics**, v.145, p. 187–196, 2016.

De Jesus, Jemmyson Romário; de Campos, Bruna Kauely; Galazzi, Rodrigo Moretto; Martinez, José Luis Capelo; Arruda, Marco Aurélio Zezzi. Bipolar disorder: recent advances and future trends in bioanalytical developments for biomarker discovery. **Analytical and Bioanalytical Chemistry**, v.407, p.661–667, 2015.

Acknowledgements

I thank God for the life.

I really thank my mother (Eliene Santos de Jesus) who is more than a mother, she is father and friend, teaching me all the principles of a good man, making no effort to get here. I dedicate this work to you. I love you so much.

I thank my fiancée Tatianny Araújo for her love, affection and, above all, her understanding in moments of absence. I LOVE YOU.

I really thank my uncles Anselmo (Nininho), Ivanilde (Nidinha), Elenilde (Elzinha), Jairo (Neguinho) and Gerson for the support and encouragement of always. Thank you!

I thank my brothers Ruan and Augusto for sharing all the moments of my life. Thank you!

I also thank my godchildren Pedro Paulo and Bernardo.

I am so grateful my supervisor Prof. Dr. Marco Aurélio Zezzi Arruda for guidance, friendship, understanding and teachings. THANK YOU VERY MUCH!

I Thank my supervisor of the master's degree Prof. Dr. Sandro Navickiene for the teachings, formation and consolidation of my academic career. I will always be grateful.

To Prof. Dr. Paulo César de Lima Nogueira for guidance during the scientific beginning, providing my commitment in the scientific field.

I am very grateful to my friends Alan, Bruno, Charlene, Darlisson, Hugo, Iara, Jandysson, Karine, Karol, Leociley, Luana, Maria de Fatima, Otávio, Paloma and Rafaelly.

I thank all the colleagues and friends who were or are part of the GEPAM group, Alejandra, Alejandro, Alisson, Amaury, Bruna, Cicero, Eraldo, Francine, Gustavo, Humberto, Isabele, Ivanilce, Katherine, Larissa, Luana, Rafael, Rodrigo, Tatiane. You were important to perform this work.

I thank all the friends of the Bioscope group, especially Ana Laço, Eduardo, Hugo López, Hugo Santos, Marta and Suzana for good moments of relaxation during my mission at Universidade Nova de Lisboa.

I am so grateful Prof. Dr. José Luis Capelo Martinez and Prof. Dr. Carlos Lodeiro for the supervision during my Ph.D. training in Portugal.

I thank CAPES for granting the Ph.D. scholarship in Brazil, as well as my mission in Portugal (process 88887.115406/2015-00).

I thank the Profs. Dr. Cláudio Muller Banzato and Dr. Ivo Raimundo Junior, as well as the Ph.D. students Rafael Fernades and Tatiani Lima for the collaboration in the development of my thesis, resulting in the publication of scientific articles.

Contents

Abbreviations

1-DE	One-dimensional gel electrophoresis
2-DE	Two-dimensional gel electrophoresis
2-D DIGE	Two-dimensional differential in-gel electrophoresis
2-D PAGE	Two-dimensional polyacrylamide gel electrophoresis
ACN	Acetonitrile
AFS	Atomic fluorescence spectrometry,
ALB	Albumin
AMYP	Pancreatic alpha-amylase
APOA1	Apolipoprotein A-I
APOA4	Apolipoprotein A-IV
APOC3	Apolipoprotein C-III
ApoD	Apolipoprotein D
APOH	Beta 2-glycoprotein1
Be	Beryllium
Bi	Bismuth
C4A	Complement C4-A
C4B	Complement C4-B
CA	Cellulose acetate
CBB	Coomassie Brilliant Blue
Cd	Cadmium
CD248	Endosialin
CE	Cellulose
Ce	Cerium
Cf	Final concentration
CH_4	Methane
CHAPS	3- [(3-Chloroamidopropyl) dimethylammonium] -1-propane, 3- [(3-cholamidopropyl) dimethylammonium] - 1-propanesulfonate
CHCA	α-cyano-4-hydroxycinnamic acid
Ci	Initial concentration

CID	Collision-induced dissociation
Co	Cobalt
Cu	Copper
Da	Dalton (1 Da = 1 g mol^{-1})
DDA	Data dependent analysis
DLS	Dynamic light scattering
DRC	Reaction cell and collision
DRX	X-ray diffraction
DTT	1,4-dithiothreitol
EGF	Pro-epidermal growth factor
ESI	Electrospray ionization
Fe	Iron
H_2SO_4	Sulfuric acids
HC	Healthy controls
HCA	Hierarchical clusters analysis
HCF	Healthy family Control
HCl	Hydrochloric acid
HCNF	Healthy non-family control
He	Helium
HF	Hydrofluoric acid
HNO_3	Nitric acid
ICP-MS	Inductively coupled plasma mass spectrometry
ICP-OES	Inductively coupled plasma optical emission spectrometry
IEF	Isoelectric focusing
IGHG1	Ig gamma-1 chain C region
IgK	Immunoglobulin kappa
In	Indian
IPG	Immobilized pH gradient
IT	Iontrap
ITIH4	Inter-alpha trypsin inhibitor heavy chain H4
LC-MS	Liquid chromatography mass spectrometry
Li	Lithium
MALDI-TOF MS	Matrix-assisted laser desorption ionization time of flight mass spectrometry
Mn	Manganese
MNPs	Magnetic nanoparticles
MS	Mass spectrometry
MS/MS	Mass Spectrometry in sequence
MW	Molecular weight
NC	Nitrocellulose
NH_3	Ammonia
NID1	Nidogen-1
O_2	Oxygen
OD	Other disorders
Pb	Lead

PBS	Saline phosphate buffer
PCA	Principal components analysis
pI	Isoelectric point
PIGR	Polymeric immunoglobulin receptor
PM	ProteoMiner®
Q	Quadrupole
RF	Radio frequency
RNAS1	Pancreatic ribonuclease
RSD	Relative standard deviation
SAMP	Amyloid serum P
SCZ	Schizophrenia
SD	Standard deviation
SDS	Sodium dodecyl sulfate
SDS-PAGE	Sodium dodecyl sulfate polyacrylamide gel electrophoresis
SEM	Scanning electron microscopy
SERPINA1	Alpha-1-antitrypsin
SPE	Solid phase extraction
TCA	Trichloroacetic acid
TEMED	N, N', N, N'-tetramethylethylenediamine
TF	Transferrin
TFA	Trifluoroacetic acid
TOF	Time of flight
TTR	Transteritine
UV	Ultraviolet
VASN	Vasorin
VTN	Vitronectin
WFDC5	Wap four-disulfide core domain protein 5
XRF	X-Ray fluorescence
Zn	Zinc

List of Figures

List of Tables

Chapter 1
Introduction

Bipolar disorder (BD) is a psychiatric disorder characterized by mood changes, ranging from mania episodes to depression [4, 12]. BD is a chronic, severe and recurrent disease that affects about 3% of the world population, being one of the main causes of death of patients by suicide [1, 7].

Considered as a complex disease due its high degree of similarity with other psychic disorders, such as schizoaffective disorder and schizophrenia, BD presents a significant lack of scientific knowledge about its pathogenesis. Some patients with BD are, for example, misdiagnosed as unipolar depressives, and, as a consequence, are submitted to inadequate and ineffective treatments [2, 3, 11]. Therefore, it is necessary to identify potential biomarkers for BD, which will help in the correct diagnosis of the disorder and for understanding of the biochemical changes associated with the disease and/or its pathophysiology, as well as in the promotion of the best treatment for the disease.

Biomarkers are considered substances indicative of normal biological processes, of pathogenic processes or of pharmacological responses to a pharmaceutical intervention [15]. In case of protein biomarkers or metal ions, specifically, they are defined as being proteins or ions that present changes in their concentration or state when associated with some stimulus or disease [13].

Human blood (serum and plasma) has been extensively studied to identify and quantify chemical species (biomolecules and metal ions), with a focus on biomarkers, which may aid in the early diagnosis, prognosis and disease progression [5, 8, 10]. However, the complexity of such samples imposes many analytical challenges [4]. For example, in proteomic studies, tens of thousands of proteins are present in blood serum samples, with a concentration range varying by 10 orders of magnitude [6]. However, of these tens of thousands of proteins present in the blood serum, only two proteins (albumin and immunoglobulins) make up about 80% of the total protein content in the serum, thus representing the proteins of greater abundance [4, 6]. Although such proteins are important for the correct functioning of the biological system, albumin and immunoglobulin may interfere in the detection of those

© Springer Nature Switzerland AG 2019
J. R. de Jesus, *Proteomic and Ionomic Study for Identification of Biomarkers in Biological Fluid Samples of Patients with Psychiatric Disorders and Healthy Individuals*, Springer Theses, https://doi.org/10.1007/978-3-030-29473-1_1

proteins of lower abundance. These proteins (smaller abundances) are generally considered as potential biological markers of human diseases [4]. In addition, another example of analytical challenge found in a biomarker study for human diseases is directly related to the ionomic strategy. In this study type, the presence of organic compounds, such as amino acids, lipids, polysaccharides, etc. may compromise the sensitivity and detection of the analytical method [9]. Thus, cleaning procedures are required to remove or reduce interfering compounds. In this sense, the sample preparation remains as a fundamental step to minimize the samples complexity, such as blood serum. Thus, the proposal of this Ph.D. thesis is to apply two omics strategies (proteomics and ionomics) widely used to found biomarkers. The hypothesis is that with the application of these strategies in a blood serum sample of patients diagnosed with BD, a panel of chemical species (proteins or metal ions) can be found and can be used to differentiate BD from other psychiatric diseases, such as schizophrenia. However, for this, different procedures of sample preparation have been evaluated for the two omic strategies. For the proteomic study, three different methods were evaluated for the removal of abundant proteins: (i) depletion using magnetic nanoparticles (MNPs); (ii) depletion using chemical agents (dithiothreitol, DTT, and acetonitrile, ACN); and, finally, a commercial technique widely used in this study type (iii) the ProteoMiner® kit. For the ionomic study, a methodology based on the decomposition of sample using high power ultrasound was developed and suggested as a method to decompose the blood serum in order to determine the metal ions of interest. To evaluate the efficiency of such a proposed method, the technique of microwave-assisted decomposition (a technique widely used in ionomic studies) was used as quality control.

To design the proteomic and ionomic profile of patients with BD, blood serum from such patients was compared to blood serum from patients diagnosed with schizophrenia (SCZ) and other mental disorders (OD), as well as compared to blood serum from healthy individuals (familial, HCF, and non-familial, HCNF).

For the ionomic study, six elements were chosen to be monitored in the serum of the interest groups: Zn, Cu, Fe, Li, Cd, and Pb. Elements were chosen because their importance as important nutrients (Zn, Cu, Fe) for the correct functioning of the biological system, especially in neurological system, as well as the direct influence of environmental contaminants (Pb and Cd), which may influence the triggering of neurological diseases.

It is important to highlight that the identification and quantification of protein and/or metal requires the application of efficient, robust and sensitive analytical techniques [14]. Thus, the combination of analytical separation techniques such as polyacrylamide gel electrophoresis (1-DE, 2-DE and 2-DIGE) with mass spectrometry techniques (ESI-QTOF-MS, MALDI TOF-MS and ICP-MS) are presented as significant strategies for such a challenge, and in this sense such techniques have also been used in this thesis.

For a better understanding of the experimental workflow, as well as a better explanation of the obtained data, this text was divided into three chapters: the first and the second refer to the proteomic and ionomic studies applied to blood serum, respectively, to identify species candidate to biomarker for BD while in the third chapter is presented the results obtained during my mission in Portugal, addressing the development of a sample preparation methodology for urine (another biological sample as important as human blood used in study of the discovery of biomarkers of human diseases).

References

1. Belzeaux R, Correard N, Boyer L, Etain B, Loffus J, Bellivier F, Bougerol T, Courtet P, Gard S, Kahn JP, Passerieux C, Leboyer M, Henry C, Azorin JM (2013) Fundamental academic centers of expertise for bipolar Disorder (FACE-BD) collaborators. Depressive residual symptoms are associated with lower adherence to medication in bipolar patients without substance use disorder: results from the face-bd cohort. J Affect Disord 151(3):1009–1015
2. Baldassano CF (2005) Assessment tools for screening and monitoring bipolar disorder. Bipolar Disord 7:8–15
3. Chudal R, Sourander A, Polo-Kantola P, Hinkka-Yli-Salomaki S, Lehti V, Sucksdorff D, Gissler M, Brown AS (2014) Perinatal factors and the risk of bipolar disorder in finland. J Affect Disord 155:75–80
4. De Jesus JR, Galazzi RM, Lima TB, Banzato CEM, Almeida Lima e Silva LF, Rosalmeida Dantas C, Gozzo FC, Arruda MAZ (2017) Simplifying the human serum proteome for discriminating patients with bipolar disorder of other psychiatry conditions. Clin Biochem 50(18):1118–1125
5. De Jesus JR, Silva Fernades R, Souza Pessôa G, Raimundo Jr IM, Arruda MAZ (2017) Depleting high-abundant and enriching low-abundant proteins in human serum: an evaluation of sample preparation methods using magnetic nanoparticle, chemical depletion and immunoaffinity techniques. Talanta 170:199–209
6. Fernández-Costa C, Reboiro-Jato M, Fdez-Riverola F, Ruiz-Romero C, Blanco FJ, Martínez JLC (2014) Sequential depletion coupled to C18 sequential extraction as a rapid tool for human serum multiple profiling. Talanta 125:189–195
7. Fulford D, Peckham AD, Johnson K, Johnson S (2014) Emotion perception and quality of life in bipolar disorder. J Affect Disord 152–154:491–497
8. Giusti L, Mantua V, Da Valle Y, Ciregia F, Ventroni T, Orsolini G, Donadio E, Giannaccini G, Mauri M, Cassano GB, Lucacchini A (2014) Search for peripheral biomarkers in patients affected by acutely psychotic bipolar disorder: a proteomic approach. Mol Biosyst 10(6):1246–1254
9. Gómez-Riza AL, Morales E, Giráidez I, Sánchez-Rodas D, Valesco A (2001) Sample treatment in chromatography-based speciation of organometallic pollutants. J Chromatogr A 938(1–2):211–224
10. Herberth M, Koethe D, Levin Y, Schwarz E, Krzyszton ND, Schoeffmann S, Ruh H, Rahmoune H, Kranaster L, Schoenborn T, Leweke MF, Guest PC, Bahn S (2011) Peripheral profiling analysis for bipolar disorder reveals markers associated with reduced cell survival. Proteomics 11(1):94–105
11. Jesus JR, de Campos BK, Galazzi RM, Martinez JL, Arruda MA (2014) Bipolar disorder: recent advances and future trends in bioanalytical developments for biomarker discovery. Anal Bioanal Chem 407:661–667
12. Rizzo LB, Costas LG, Mansur RB, Swardfager W, Belangero SI, Grassi-Oliveira R, Mcintyre RS, Bauer ME, Brietzke E (2014) The theory of bipolar disorder as an illness of accelerated aging: implications for clinical care and research. Neurosci Biobehav Rev 42:157–169

13. Sharma S, Moon CS, Khogali A, Haidous A, Chabenne A, Oio C, Jelebinkov M, Kurdi Y, Ebadi M (2013) Biomarkers in parkinson's disease (recent update). Neurochem Int 63(3):201–229
14. Timerbaev A, Pawlak K, Gabbiani C, Messori L (2011) Recent progress in the application of analytical techniques to anticancer metallodrug proteomics. Trends Anal Chem 30(7):1120–1138
15. Von thun und hohenstein-blaul N, Funke S, Grus FH (2013) Tears as a source of biomarkers for ocular and systemic diseases. Exp Eye Res 117:126–137

Chapter 2
Application of Proteomic Strategy for the Identification of Differential Proteins Candidates to Biomarkers of Bipolar Disorder

2.1 Objectives

2.1.1 General Objective

Evaluate the proteomic profile of the serum of individuals with BD to identify differential proteins with the potential to be biomarkers of this disease. For this, a comparative study using blood serum from individuals with BD, healthy individuals and individuals diagnosed with other psychiatric disorders, such as schizophrenia, was performed.

2.1.2 Specific Objectives

- To evaluate different sample preparation procedures to simplify the sample complexity, removing proteins of high abundance, and enriching those proteins of lower abundance;
- To separate the blood serum proteins from the volunteers present in the study groups using two-dimensional polyacrylamide gel electrophoresis (2-DE), in order to obtain the proteomic profiles;
- To evaluate possible differences in the expression of the proteins between the samples evaluated using the two-dimensional differential electrophoresis in polyacrylamide gel (2-D DIGE) technique;
- To identify the proteins that exhibited differences in expression using mass spectrometry (nLC-ESI-QTOF MS) and Mascot program;
- To establish possible relationships between differentially abundant proteins and the pathophysiology of BD.

© Springer Nature Switzerland AG 2019   5
J. R. de Jesus, *Proteomic and Ionomic Study for Identification of Biomarkers in Biological Fluid Samples of Patients with Psychiatric Disorders and Healthy Individuals*, Springer Theses, https://doi.org/10.1007/978-3-030-29473-1_2

2.2   Review of the Literature

Overview presented below will focus on the main issues addressed in this chapter.

2.2.1   Bipolar Disorder

Bipolar disorder (BD) is a chronic and recurrent disease with high levels of morbidity and mortality, and is considered to be the fifth major cause of disability among all diseases [15]. BD is characterized by period of mania, depression, hypomania or mixed episodes, with euthymic periods (stable mood state) [27].

In an episode of mania, the individual has symptoms, such as low need for sleep, impulsive behavior and an increase in the practice of activities. This episode can last about a week. In episodes of depression, which can last two or more weeks, the patient has a loss of interest in performing physic exercise, reduced energy, and presents ideas or attempts at suicides. Hypomania presents the same symptoms of manic episodes, however with shorter periods. Mixed episodes are characterized by alternations between mania and depression [43].

When diagnosed with BD, patients are treated with traditional medicines. Among the most common medications, lithium therapy (a mood stabilizer), referred to more than 60 years in the clinical setting, stands out. Other medicines, such as chlor-promazine and olanzapine (antipsychotics), are also considered as pharmacological treatment options for BD [32] (Sussulini 2010). It is worth noting that the neurologi-cal action mechanisms of these medicines in BD are still unknown and/or unclear [44, 45]. This fact justifies the deepening of new studies in search of biomarkers for bipo-lar disorder. The treatment of BD is divided into three phases: (i) acute phase (treats mania without causing depression and/or consistently improves depression without causing mania); (ii) continuation phase (aims to stabilize the benefits, reduce the side effects and reduce the possibility of relapse); (iii) maintenance phase (aims to prevent mania and or depression, maximizing functional recovery) [19].

Some studies suggesting the existence of genetic factors in the susceptibility to bipolar disorder, both in terms of family components and also in associations with other psychiatric disorders, with schizophrenia being the main one [4, 16, 28].

However, the complexity, heterogeneity and absence of BD biomarkers are factors that may contribute to the disease misdiagnosis. Recent literature reports the diffi-culty of differentiating BD, unipolar depression, schizophrenia and schizoaffective diseases [38, 39]. The high prevalence of depression in relation to hypomanic symp-toms over BD and the presence of manic symptoms during depressive episodes are some of the main reasons for the difficulty of differentiating BD and other psychoses [6].

Currently, the BD diagnosis is made exclusively using different types of ques-tionnaires established by World Health Organization (WHO) [8]. However, like any other disease with physiological changes in the body, patients with BD can to present differences in their biological system when compared to healthy individuals. In this sense, the discovery of such differences can be important to help in the correct diagnosis of BD. Thus, the discovery of differential proteins candidates for biomarkers appears as an important challenge for the diagnosis and understanding of biochemical alterations associated with this disease.

2.2.2 Biomarkers

Biomarkers discovery focused on diagnosis is an important tool to understand possible abnormalities or failures in the human body [48]. A biomarker is defined as a substance that can be measured and evaluated, allowing to indicate biological normal processes, pathogenic processes or pharmacological responses to a therapeutic intervention. Different markers can be used to perform different evaluations, for example:

- To diagnose a disease;
- To predict the natural outcome of an individual with a particular disease;
- To predict whether the individual will benefit from a specific treatment;
- And, to evaluate an individual's response to a particular treatment.

If founded, the biomarkers may help to apply the best treatment in curing or combating the symptoms of a disease [2].

The field of discovery, development and application of biomarkers has been the subject of intense interest and activity, especially with the emergence of new omic technologies, such as proteomics.

Proteomics is a science field that examines all the proteins involved in a cell or tissue of an organism. The proteins assessment present in an organism at a given time is essential to understand the biological processes that occur in healthy and diseased organisms. In recent years, many efforts have been employed to identify differential proteins with potential for biological markers of psychological disorders, such as BD, schizophrenia, depression, among others [21, 35]. For this, different types of samples have been explored, for example, body fluids (blood, urine and saliva) [42] and brain tissues [33]. Human blood is the most used sample in the proteomic analysis for the discovery of biomarkers [35]. The choice of human blood as the primary study sample is directly related to its ability to provide complete information of the biological system as well as it is the main sample used in clinical trials.

However, although human blood is the primary source for the search for biological markers of human disease, such a sample presents important analytical challenges because of its complexity. Thus, to overcome such challenges, techniques based on separation, quantification and identification, presenting high sensitivity and selectivity as well as efficient sample preparation strategies is required. Thus, in the following sections, an overview regarding sample preparation procedures and analytical techniques used in the search for protein differences is presented.

2.2.3 Sample Preparation Procedures to Simplify the Proteome of Biological Samples

In proteomics, the search for biomarkers in biological fluids, such as human blood present many analytical challenges, mainly due to the high amount of proteins present in this type of samples. For example, the levels of protein concentration in blood (plasma and serum) can vary by more than ten orders of magnitude. In this sense, proteins with higher concentrations may interfere in the detection of those proteins of lower abundance, making it difficult to identify biomarkers. To solve this challenge, several methods have been presented in the literature as a way of removing or minimizing the presence of such abundant proteins. For example, according to Javanmard et al. [23], the most abundant proteins can be depleted using an on-chip automated platform [6] composed of two components. The first component consists of a microfluidic mixer for a mixture between antibodies and blood samples (consisting of cells and serum proteins), which is injected into a second component, a filtering trench, which captures all cells. The cells retention and beads that captured the highly abundant proteins is made so that they are pushed into the filter trench by applying a negative dielectrophoretic force, and only those proteins of low abundance will flow, avoiding equipment, as for example, centrifuges and columns for the proteins purification.

Another example of abundant protein depletion of human blood is the use of commercial kits, and in this sense the ProteoMiner kit stands out. The ProteoMiner kit (BioRad) is based on the selective adsorption of proteins in a stationary phase containing a set of impregnated peptides. Such adsorption occurs under limited binding conditions. Thus, the proteins of low and high abundance are enriched and reduced simultaneously, allowing the detection of proteins of low abundance. The ProteoMiner kit was successfully applied in the search for BD biomarkers, finding apolipoprotein A-I as a biomarker candidate protein in response to lithium treatment performed by patients with BD [44, 45].

In addition to these strategies, other alternative procedures have also been employed to deplete high-abundance proteins [14, 24, 41], such as immunoaffinity chromatographic columns containing immobilized antibodies as well as the use of chemical reagents (acetonitrile, ACN and dithiothreitol DTT) [25, 49] and the use of magnetic nanoparticles (MNPs) [1].

2.2.4 Analytical Techniques Applied in the Separation and Identification of Differential Proteins

2.2.4.1 Gel Electrophoresis Techniques

One- or two-dimensional gel electrophoresis (1-DE or 2-DE) has been widely used in clinical proteomics studies. This method allows evaluating, with significant resolution, proteins from complex samples, such as blood serum [37]. One-dimensional electrophoresis (1-DE), also known as sodium dodecyl sulfate polyacrylamide gel electrophoresis (SDS-PAGE) is a popular method widely used in (i) proteomic comparison in different samples, (ii) as a second dimension in 2-DE analysis and (iii) in a western blotting proteomic validation procedure. In two-dimensional gel electrophoresis (2-DE), the proteins are separated according to two basic principles of separation: (i) first separation according to their isoelectric point (pI), at a pH gradient in polyacrylamide gel, and (ii) second-dimensional electrophoresis using sodium dodecyl sulfate (SDS) as a denaturant for the separation of the proteins in polyacrylamide gel according to their molecular weight [29] (Fig. 2.1).

Two-dimentional electrophoreses technique has the advantage of simplicity and selectivity in separating hundreds of proteins present in complex biological samples, as well as the facility of staining, allowing the quantification of a larger number

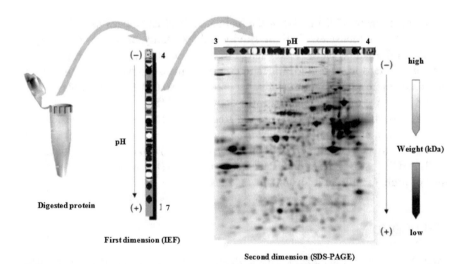

Fig. 2.1 Representative image of 2-DE technique. For the first dimension, the proteins in solution are separated according to their pI by isoelectric focusing (IEF) applying an electric field. The electric force moves the proteins (charges) until it reaches the corresponding pI. The strip is loaded onto a one-dimensional gel (SDS-PAGE), and a new electric field is applied. This time, the proteins separate according to the corresponding sizes

of proteins. However, the 2-DE technique has some limitation, as for example, non-quantification of proteins with high pI and the non-recovery of hydrophobic and high molecular weight proteins [37].

The 2-D DIGE technique, on the other hand, also constitutes another analytical technique applied in comparative proteomic studies [5]. This technique is based on the use of fluorescent dyes, which allow performing quantitative proteomic comparisons between two samples, which are solved in the same gel. In addition, the high sensitivity of these dyes is remarkable, which allows the detection of low-abundance proteins, when these dyes are compared to others used in the detection of protein spots. In this technique, complex protein mixtures are labeled with fluorescent dyes prior to electrophoretic separation, allowing the detection and quantification of proteins with different concentrations in a single gel. The dyes are excited at specific wavelengths so that the fluorescence intensity compared allows quantification of each protein present in the gel [47] (Fig. 2.2).

To obtain the data of the 2-D DIGE technique, high resolution image pickers and computers associated to the method are used. Such equipment enables fast data exploration with spot detection, normalization, quantification, protein profiling and reporting for exporting data.

2.2.4.2 Mass Spectrometry

Mass spectrometry (MS) is one of the main analytical techniques used for protein analysis due its robustness, high sensitivity and specificity, as well as its ability to generate information on the total content of complex samples in a short time [11]. One of the steps in protein identification experiments using MS is to fragment the protein of interest into a representative set of peptides so that they have molar masses within the range of detectable mass by spectrometers [45].

Mass spectrometers consist of an ion source, a mass analyzer, an ion detector, and a data acquisition unit. The most used ionization techniques in protein analysis are matrix assisted laser desorption/ionization (MALDI) and electrospray ionization (ESI) and the basic types of mass analyzers are iontrap (IT), quadrupole (Q) and the time of flight (TOF), orbitrap and others [12, 46].

MALDI-Q/TOF

MALDI technique is an ionization method capable of analyzing high molecular mass, non-volatile and thermolabile compounds such as peptides, synthetic polymers and organic compounds in a wide range of molar masses [31].

The technique uses the beam of a laser to supply energies to the molecules, providing their desorption and leading them to the gas phase in its ionic form [12].

In analysis procedure, the sample is first subjected to a proteolytic digestion (generally using trypsin), after that formed peptides are then mixed with excess acidic organic matrix on a MALDI plate, where both crystallize after solvent evaporation.

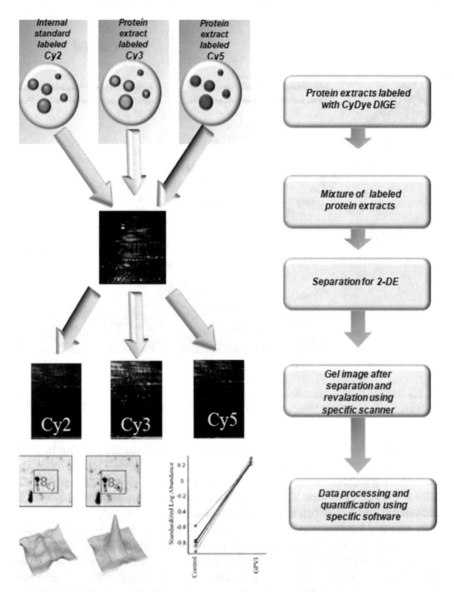

Fig. 2.2 Representative image of 2-D DIGE technique. Proteins extracted from control and patient samples are labeled with different CyDyes DIGE dyes, mixed and separated by 2-DE. The gels are scanned using an instrument that detecting each CyDye independently. The images are analyzed and the spots are determined corresponding to the same proteins with different levels of abundance. After 2D-DIGE, the protein may be identified by mass spectrometry

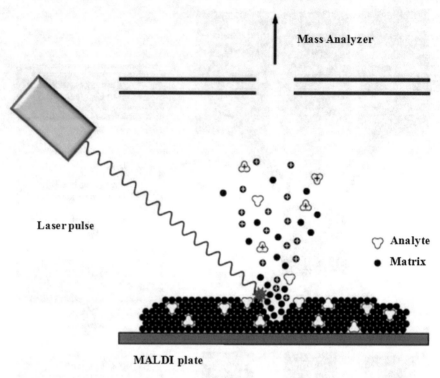

Fig. 2.3 Lustration of an ionization source by MALDI. In this type of ionization, the peptides are crystallized with an organic matrix and, after laser bombardment, they are ionized

Under high vacuum, the mixture is subjected to short pulses of laser that impinge on the crystallized sample, causing its sublimation. The region irradiated by the laser is heated and causes ions desorption, being transferred through electromagnetic fields to the mass analyzer. The generated ions will be separated according to their mass/charge (m/z) ratio before reaching the detector [12] (Fig. 2.3).

In the mass spectrometer, the determination of m/z can be performed through a TOF analyzer, in which the ions are separated and analyzed along a tube of specific length according to their different velocities. This analyzer can be combined with another to increase efficiency in the determination of ions, for example, many studies report the combination of TOF with the quadrupole (Q) analyzer. In this case, the technique is called sequential mass spectrometry, in which specific ions are selected and subjected to fragmentation to obtain a spectrum of precursor ions. In the quadrupole, ions are selected by applying radiofrequency (RF) voltages on the bars, causing inversion of polarity between them. Thus, the quadrupole analyzer acts as a filter, allowing only the most stable ions in this field to be selected [26].

In this way, in the sequential module (Q/TOF), the ions are initially selected in quadrupole, fragmented in a collision cell and the masses of the fragmented ions are determined by the TOF analyzer. Thus, a characteristic spectrum of the analyzed

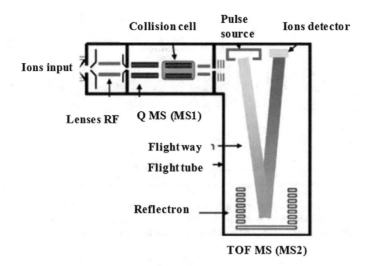

Fig. 2.4 Representative image of sequential system (Q/TOF). After ionization of the peptides, the ions of interest are selected in quadrupole (Q), fragmented in the collision cell, and after that, the generated fragments are separated by time of flight based on the molar mass of each

protein is obtained. This mass spectrum is compared to a database containing known protein sequences. This comparison allows the identification of the protein under analysis [26] (Fig. 2.4).

Rico Santana et al. [40] reported the search for new biological markers for the early diagnosis of ischemic stroke (AVC) using MALDI-TOF-MS as a screening technique to obtain the mass spectra of the extracted peptides. In this study, the researchers collected 63 blood serum samples from patients with neurological diseases, 45 of whom had ischemic stroke and 18 patients with other neurological disorders. Fifty and six serum samples from healthy subjects were analyzed and considered as control samples. In the end, no biomarker was identified to differentiate between patients with ischemic stroke and another neurological disease. However, they were able to differentiate patients with ischemic stroke from healthy individuals.

ESI-Q/TOF

ESI ionization is another technique also widely used for protein identification [46]. This technique allows transferring the solution ions to the gas phase. This versatility significantly increases the amount of substances that can be determined. Such ionization involves the formation of an electrolytic spray, which generates small droplets from which the ions are released [34] (Fig. 2.5).

In this technique, a high voltage source (1000–7000 V) is required. It is in contact with the solution containing the electrolytes. This solution is pumped through a microcapillary (i.e. 50–100 μm) at a flow rate of less than 10 μL min^{-1}, in the

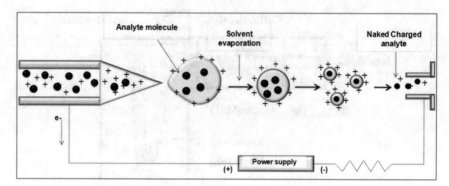

Fig. 2.5 Representative image of ionization source by ESI. In this ionization technique, the aqueous solution containing the analyte is injected through a capillary, forming an aerosol of highly charged droplets which upon evaporation of the solvent generate ionized forms of the analyte

case of a flow less than 1 μL min^{-1}, the process is called nanoelectrospray. When a positive potential is applied in the solution, the positive ions moves away towards the less positive region that is towards the counter electrode. Therefore, the formed drop at the capillary tip will be enriched in positive ions. As the charge density increases in the drop, the electric field formed between the capillary and the counter electrode increases causing the drop deformation. The frequency of this process depends on the electric field magnitude, the surface solvent tension and the solution conductivity [34].

The subsequent step is the obtaining of the mass spectra, through the same techniques described in the previous item, by the sequential analysis, of type Q/TOF.

In a study aimed at differentiating patients with BD and schizophrenia, Iavarone et al. [22] have suggested the use of saliva to evaluate proteins and peptides. In this study, nLC-ESI-MS/MS was applied to identify and quantify the relative peptides. The saliva samples were treated with 0.2% (v/v) trifluoroacetic acid and centrifuged at 9000g for 5 min. All proteins and peptides identified were involved in the immunological process of both disorders. The proteins identified showed an increase in abundance with factor of more than 10 times when compared to the control group. Levels of α-defensins, for example, were significantly altered, suggesting that their differential concentrations could be associated with increased neutrophil activity and/or the contribution of other cells related to adaptive immunity. This study also suggests that dysregulation of the peripheral white blood cell immune pathway may be associated with schizophrenia, opening up new possibilities for research to elucidate the mechanisms of activation in mood disorders.

2.3 Experimental

2.3.1 Materials and Reagents

2.3.1.1 Equipment

- Analytical balance (Shimadzu, model AX200, class I);
- Direct current source (Armersham Biosciences, Sweden);
- Scanner, Image ScannerTM II model (GE Healthcare, Sweden);
- Model Scanner EttanTMDIGE Image (GE Healthcare, Sweden);
- Mili-Q purification system, model Quantum TM cartridge (Millipore, France);
- Isoelectric focusing system, Ettan TM IPGphorII TM model (GE Healthcare, Sweden);
- SDS-PAGE electrophoresis system, model SE 600 Ruby (GE Healthcare, Sweden);
- Ultracentrifuge, model Bio-Spin-R (BioAgency, Brazil);
- Analytical purity reagents from Amersham Biosciences (Sweden), BioAgency (Brazil), J.T. Baker (USA) and Merck (Germany), and glassware for routine use in the field of bioanalytics.

2.3.1.2 Buffers and Solutions

- Agarose solution: 0.8% agarose (w/w), bromophenol blue 0.002% (w/v), run buffer diluted 3 times in deionized water;
- Fixing solution: 40% (v/v) ethanol, glacial acetic acid 10% (v/v), deionized water;
- Bleaching solution: 5% (v/v) methanol, 7.5% (v/v) glacial acetic acid, deionized water;
- Dye solution (colloidal coomassie): 10% (v/v) phosphoric acid, 10% ammonium sulfate (w/v), comassie 0.12% G-250, 20% (v/v) methanol, deionized water;
- Tris-HCl buffer pH 8.8: deionized water, Tris-base 1.5 mol L^{-1}, pH adjustment with hydrochloric acid (HCl);
- Rehydration buffer: urea (7 mol L^{-1}), thiourea (2 mol L^{-1}), CHAPS 2% (w/v), IPG buffer (ampholytes) pH 4–7 0.5% (v/v), blue of bromophenol 0.002% (w/v), deionized water
- Sample Buffer: Tris-HCl pH 6.8 (40 mmol L^{-1}), 10% (v/v) β-mercaptoethanol, 50% (v/v) glycerol, 10% SDS (w/v), bromophenol 0.1% (w/v);
- Buffer (10-fold concentrate): Tris-base (2 mmol L^{-1}), glycine (192 mmol L^{-1}), 0.1% SDS (w/v), deionized water;
- Gel solution: 30% (w/v) acrylamide, 0.8% (w/v) N,N-methylenebisacrylamide, deionized water.

2.3.1.3   Other Materials

- BD Vacutainer® Tubes (BD SST® II Advance)
- 2-D Clean-up Kit (GE Healthacare, Sweden)
- Immobilized polyacrylamide gel tape containing pH gradient (4–7) (130 × 3 × 0.5 mm) (Immobile™ Dry Strip, GE Healthcare, Sweden).

2.3.2   Methodology

2.3.2.1   Acquisition of Samples

This study was approved by the Research Ethics Committee of the University of Campinas (UNICAMP), and is registered under protocol number 775/2010. The patients gave their written consent prior to sample collection (model of consent form and questionnaire are attached). All patients were treatment at the psychiatric clinic of the University hospital and were diagnosed based on clinical criteria according to the International Classification of Diseases, 10th Edition (ICD-10).

Fifty-three serum samples were collected and classified into five groups: the control group was selected based on the exclusion of BD history and consists of samples of (i) healthy family control (HCF) and (ii) healthy non-family control (HCNF); (iii) patients with schizophrenia (SCZ); (iv) patients with BD; and (v) patients with other mental disorders (OD). To better understand the group's choices, it is important to note that in previous studies of our group [44, 45], differences in the proteomic and ionomic profile of patients with BD using lithium and using other medicines were evaluated. Thus, it was decided to compare patients with BD with other mental disorders, such as SCZ, and also with another group of patients presenting with other psychic disorders (OD group). Patients with other medical conditions such as cancer, AIDS, endocrinological and metabolic diseases were excluded in the recruitment process. Table 2.1 summarizes the sample characteristics.

2.3.2.2   Sample Preparation

Three different strategies were evaluated to simplify the proteome of human blood serum: (i) using chemical reagents (dithiothreitol, DTT, and acetonitrile, ACN); (ii) using immunoaffinity technique (commercial kit ProteoMiner®, PM) and (iii) using magnetic nanoparticles. The procedures are described below.

Table 2.1 Information of the volunteers involved in the research

	HCF (n = 3)	HCNF (n = 9)	BD (n = 14)	SCZ (n = 23)	OD (n = 4)
Sex (male/female)	1/2	2/7	5/9	17/6	3/1
Age (year ± SD)	39 ± 16	35 ± 8	36 ± 9	34 ± 9	31 ± 5
Diagnostic time (year ± SD)	–	–	5 ± 4	9 ± 8	5 ± 3
Smoker (male/female)	0/1	0/1	2/2	5/0	1/0

SD Standard deviation

Depletion Abundant Protein Using DTT and ACN

In this procedure, two protocols previously published using DTT [49] and ACN [25] were performed in sequence. Briefly, three aliquots of 20 μL of human serum were mixed with 2 μL of DTT (500 mmol L^{-1}), solubilised in ammonium bicarbonate (Ambic, 12.5 mmol L^{-1}) and homogenized. Samples were incubated for 60 min at 37 °C, centrifuged at 13,000g for 40 min and the supernatants transferred to a Lobind microtube (Eppendorf, Cambridge, UK). Each aliquot was then diluted with 45 μl of water and subjected to stirring. Subsequently, 85 μl of ACN was added and incubated for 10 min in an ultrasound bath. The protein precipitate was separated by centrifugation at 13,000g for 10 min at room temperature. The supernatants obtained were evaporated to dryness in a vacuum concentrating centrifuge without heating. The sample was reconstituted in 100 μL of a solution containing 8 mol L^{-1} urea, 2% (m/v) CHAPS, 3 mol L^{-1} thiourea and stored at -20 °C until quantification and subsequent electrophoresis analysis in gel.

Equalization of Proteins Using ProteoMiner® (PM)

Enrichment of low abundance proteins as well as the removal of high abundance proteins from the serum samples were performed using the PM kit according to the manufacturer's instructions. Briefly, the PM spin columns were first washed with water (2× for 5 min) and then washed (150 mmol L^{-1} NaCl, 10 mmol L^{-1} NaH$_2$PO$_4$, pH 7.4) (2× for 5 min). Four aliquots of human serum (200 μL) were then applied to the column and incubated for 2 h. Subsequently, the column was washed with wash buffer (3 × for 5 min). Protein elution was performed three times for 15 min with 100 μl elution reagent (containing urea, CHAPS and acetic acid). The eluted fractions were adjusted to neutral pH with a solution of Tris 1 mol L^{-1}, and stored at -20 °C until quantification and subsequent analysis by gel electrophoresis.

Abundant Protein Removal Using Magnetic Nanoparticles (MNPs)

The magnetic nanoparticles were acquired in collaboration with Prof. Dr. Ivo Milton Raimundo Jr. (Institute of Chemistry/Unicamp). The MNPs synthesis was performed according to the protocol of Lu et al. [30].

The depletion of high abundance proteins from serum samples using MNPs was investigated, evaluating some parameters: (i) the influence of nanoparticle size; (ii) incubation time; (iii) mass ratio of MNPs/protein; and (iv) pH of the medium in the depletion process. For this study, agitation (300 rpm) and an incubation temperature (25 °C) were predefined experimental conditions.

pH effect of the sample in the depletion procedure was evaluated varying from pH 5.5 to 8.5 using Tris-HCl solution (0.1 mol L^{-1}). Serum volumes (with a total protein concentration of 54 μg L^{-1}, determined using the 2-D Kuant kit) ranging from 25 to 90 μl were diluted to a final volume of 100 μl using phosphate buffered saline (2% v/v, pH 7.4) or Tris-HCl (0.1 mol L^{-1}, pH 5.5 or 8.5) and mixed with MNPs (5 μg) in order to obtain the following MNP/protein ratios: 1:2, 1:5, 1:10. In addition, the extraction time, which ranged from 30 to 120 min; and the size of MNPs, which ranged from 30 and 70 nm, were also investigated. After the depletion process, the supernatant and the solid phase (abundant proteins adsorbed on the surface of the MNPs) were magnetically separated. The supernatant was then transferred to a clean LoBind tube and evaporated to dryness in a vacuum concentrator centrifuge without heating. The sample was reconstituted in 100 μL of a solution containing urea (8 mol L^{-1}), CHAPS (2% w/v), thiourea (3 mol L^{-1}) and stored at −20 °C until quantification and subsequent analysis by gel electrophoresis.

2.3.2.3 Proteins Quantification

After performing the depletion procedures, the protein concentration was determined using the commercial kit for quantification (2-D Quant, GE, Healthcare). Quantification was performed following the manufacturer's instructions. Briefly, an analytical curve of the bovine serum albumin (BSA) standard was constructed, with concentrations ranging from 0.0 to 50.0 μg mL^{-1}. The calibration curve was then used to determine the protein concentration in the samples. Each experiment was performed in triplicate for quality control and statistical analysis. The following equation describes the depletion efficiency (E):

$$E = (Cf/Ci) \times 100\%$$

where Cf is the protein concentration after the depletion procedure (μg $μL^{-1}$) and Ci is the initial concentration of the protein (μg $μL^{-1}$).

Table 2.2 Scheme for 2-D DIGE analysis between the study groups

Experiment	Groups compared
1	Healthy non-familiar control versus bipolar disorder
2	Bipolar disorder versus other disorders
3	Healthy non-familiar control versus healthy familiar control
4	Healthy familiar control versus bipolar disorder
5	Schizophrenia versus bipolar disorder

2.3.2.4 Electrophoreses Analysis

SDS-PAGE

For the analyzes employing SDS-PAGE, 30 µg of protein (about 10 µl solution) of serum sample without prior treatment and with depletion procedure were mixed with sample buffer [10% (w/v) sodium dodecyl sulfate (SDS), 40 mmol L^{-1} Tris (pH 6.8), 50% (v/v) glycerol, 0.1% (v/v) bromophenol blue, 10% (v/v) mercaptoethanol] in a ratio of 1:1. Heating at 98 °C for 3 min resulted in denaturation of the sample, which was then loaded in polyacrylamide gel 12% (w/v). The electrophoretic separation was performed as follows: 200 V for 50 min and the gel stained with comassiee blue (CBB). For dye removal, the gels were incubated with a decolorizing solution (40% (v/v) methanol, 10% (v/v) acetic acid and 50% (v/v) water for 24 h at room temperature. The gels were then rinsed with water and incubated for at least 1 h. Images were acquired using ImageScanner (GE Healtchare).

2-D DIGE and Statistics

Five groups were used for the analysis of 2-D DIGE (Table 2.2). Gels from each group were prepared in triplicate. 2-D DIGE analysis was performed using Cy2 dye as the internal standard with equal amounts of the two samples under analysis. Samples from each group were pooled and labeled with Cy3 or Cy5. Protein precipitate was obtained using 20 µl Tris (1 mol L^{-1}), 480 µL cold methanol, 240 µL cold chloroform and 360 µL water. The precipitate was collected and solubilized in 40 µl lysis buffer (7 mol L^{-1} urea, 2 mol L^{-1} thiourea, 4% (w/v) CHAPS and 20 mmol L^{-1} Tris pH 8,8). Total protein determination was performed using the 2D Quant kit (GE Healthcare, Uppsala, Sweden) according to the manufacturer's instructions. For labeling with CyDye DIGE dyes, 60 µg of protein from each sample were stained with 400 pmol dyes from CyDye DIGE Fluor (GE Healthcare) and incubated on ice in the absence of light for 30 min. The labeling reactions were terminated by the addition of 1 µl lysine 10 mmol L^{-1} and incubated for 10 min. Isoelectric focusing (IFE) was performed using containing pH gradients ranging from 4 to 7 (Immobiline Dry Strips, pH 4–7, 13 cm, GE Healthcare) with a programming of 14.600 Vh. The gels were scanned with an Ettan DIGE Imager Scanner (GE Healthcare) and analyzed for proteome differences, spot detection, gels ratio and normalization based on the combined internal standard prior to quantification (Cy2). Differential abundance was considered to be statistically significant after the Student t-test if the adjustment factor was greater than 2, and $p \leq 0.05$. These values are set by the HUPO (Human Proteome Organization). After that, 2-D PAGE gels were stained with comassie blue prior to mass spectrometry analysis for the identification of the protein.

2.3.2.5 In-Gel Tryptic Digestion

For protein digestion using trypsin, the spots of interest were manually cut from the gel and placed on a microplate containing peptide affinity resin using the Montage® In-Gel digestZP kit (Millipore, Bedford, U.S.A.). The protocol of digestion and vacuum elution were performed according to de Jesus et al. [8, 9]. Briefly, CBB dye removal with 1:1 ammonium bicarbonate/acetonitrile was performed. The proteins were subjected to reduction and alkylation using DTT and ACN. Tryptic digestion was performed with approximately 166 ng of enzyme for each spot. Finally, the purified peptides were eluted from the resin using 100 μL of 0.1% (v/v) trifluoroacetic (TFA) in 50% (v/v) acetonitrile solution. For vacuum elution, a Multiscreen® Vacuum Manifold (Millipore) was used.

2.3.2.6 Protein Identification Using NanoLC-QTOF MS

LC-MS/MS analyzes of the peptides resulting from protein digestion were performed on a Waters nanoACQUITY UPLC chromatograph coupled to the Waters SYNAPT HDMS (QTOF) spectrometer equipped with a nano-ESI source. A volume of 5 μL of sample was injected into the UPLC system using a protection column (Waters Symmetry C18, 20 mm × 180 μm). The peptides were separated on a Waters BEH130 C18 analytical column, 100 mm × 100 μm and eluted at 0.65 μL min^{-1} using a linear gradient ranging from 90 to 99% (v/v) acetonitrile with formic acid (1–10%, v/v). The detection of the peptides was performed by the Waters SYNAPT HDMS spectrometer, configured to operate in the Data Dependent Analysis (DDA) acquisition mode, with the equipment acquiring one spectrum per second. When species with multiple charges are detected, the three most intense species are fragmented in the collision cell. The collision energy is selected according to the precursor m/z and the charge. spectrum acquisition was performed in positive mode using MassLynx v.4.1 software. All mass spectra were processed using Mascot Distiller (Matrix Science, London, UK) and screened against the NCBI data. The level of significance was set at $p < 0.05$, which corresponds to a score of 30.

2.4 Results and Discussion

2.4.1 Optimization of Parameters for Removal of Abundant Proteins in Blood Serum Using MNPs

In recent years, MNPs have attracted much attention as an important strategy for the removal of abundant proteins in biological samples because they provide simple, inexpensive and robust applications in the discovery of differential proteins candidate to biomarker of human diseases [50]. However, although such a strategy is convenient and rapid in the depletion process, it presents a large number of variables that may influence the adsorption efficiency of proteins on the surface of MNPs [1, 50]. For this reason, four parameters were evaluated in triplicate for quality control and statistical analysis: (i) size of NPMs; (ii) incubation time; (iii) mass ratio of MNP/mass of protein; and (iv) medium pH. The stirring and the incubation temperature were previously set at 300 rpm and 25 °C, respectively.

To assess the influence of nanoparticle size on the depletion of abundant proteins in blood serum, two different sizes (30 and 70 nm) of MNPs were evaluated. Figure 2.6 shows the characterization of MNPs using scanning electron microscopy (SEM), dynamic light distribution (DLS), and thermogravimetric analysis.

Experiment was performed using an initial ratio of 1:2 (MNP mass/protein mass), initial pH of 7.4 and stirring for 30 min. After 30 min incubation, the supernatant and the MNPs containing the adsorbed proteins were then magnetically separated. The remaining protein in the supernatant was quantified. The results are shown in Fig. 2.7 (panel a). From Fig. 2.7 (panel a) it can be seen that both nanoparticles (30 and 70 nm) showed efficiency in the removal of the protein from the serum sample as compared to the crude serum. However, although the 70 nm MNPs had lower surface area, they curiously provided better results in terms of protein depletion, reaching a protein recovery of approximately $4 \, \mu g \, \mu L^{-1}$ (Fig. 2.7, panel a, bar graph), while the 30 nm MNPs achieved recoveries of approximately $7 \, \mu g \, \mu L^{-1}$ (Fig. 2.7, panel a, bar graph). These results can be explained in terms of MNPs agglomeration. In other words, it is suggested that MNPs of 30 nm may have undergone rapid in situ agglomeration, influencing the low adsorption of proteins. However, for MNPs of 70 nm, such in situ agglomeration was minimal. Thus, this fact indicates that the aggregation of MNPs may affect the effective interaction between proteins and MNPs. As MNPs of 70 nm showed significant removal of the abundant proteins, they were therefore considered for the next experiments.

Another important parameter that may affect the capacity and kinetics of protein adsorption on the surface of nanoparticles is protein concentration [3]. To investigate the influence of the ratio MNPs mass/protein mass on the depletion process, a series of blood serum aliquots were mixed with MNPs (70 nm), resulting in proportions (MNP mass/protein mass) of 1:2, 1:5 and 1:10. Figure 2.7, panel b, shows a decrease in depletion efficiency, with an increase in protein concentration. This fact can be explained by the system saturation. In other words, when the amount of adsorbent (i.e. MNPs mass) is kept constant, and the protein concentration is increase in the

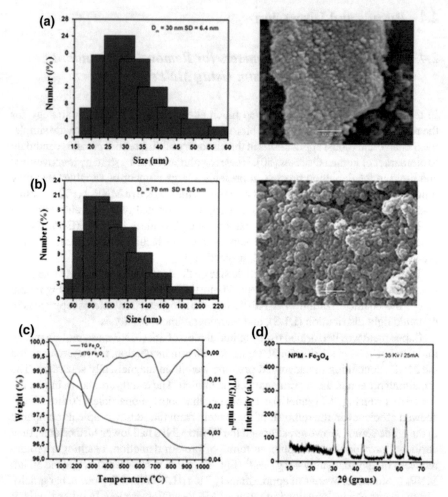

Fig. 2.6 Scanning electron microscopy (SEM) images and size distribution of Fe_3O_4 nanoparticles. **a** 30 nm; **b** 70 nm; **c** thermogravimetric analysis (TG/dTG) of Fe_3O_4 particles and **d** X-ray diffraction (XRD) spectrum for Fe_3O_4

system, a decrease in the available adsorption sites is reduced, consequently leading to a reduction in efficiency of protein removal [3]. As the ratio 1:2 (MNP/protein) presented better results in relation to the proteins removal higher abundance, such proportion was then selected for the other experiments.

The effect of the incubation time on the protein depletion process was also evaluated, and the stirring time varied between 30, 60 and 120 min. From Fig. 2.7 (panel c), it can be seen that there is no significant change in depletion efficiency with rated agitation time. Thus, 30 min of agitation was sufficient to ensure a significant removal of the abundant proteins, and, therefore, such time was selected as the ideal incubation time. To assess the influence of pH on abundant protein removal from human

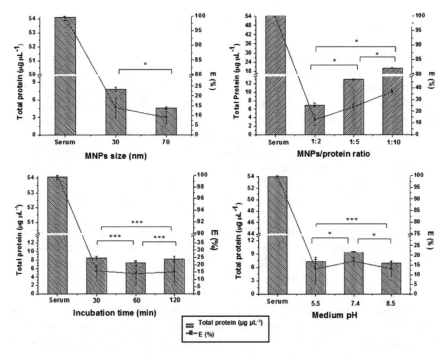

Fig. 2.7 Evaluation of the parameters efficiency used to remove abundant proteins. **a** Size MNPs effect; **b** MNPs mass/protein mass influence; **c** incubation time influence; **d** pH effect. Results obtained from three independent depletion experiments. The square curve plot represents the recovery of the resulting protein (%) after depletion. The bar graph represents the recovery of the protein in $\mu g\ \mu L^{-1}$. The statistical analysis was performed with t-test and considered significant for $*p < 0.001$ and $***p > 0.05$

serum, 5 μg MNPs (70 nm) were added to 25 μL serum diluted in buffers with 75 μL Tris-HCl at different pH (5.5 or 8.5) or phosphate buffered saline (PBS, pH 7.4). The samples were then shaken at 300 rpm for 30 min at 25 °C. According to Fig. 2.7, panel d, the lowest adsorption capacity was obtained at pH 7.4, while the highest values were observed at pH 5.5 and 8.5. This observation can be explained by the maximum binding capacity between MNPs and proteins reached at the pH near the isoelectric point (pI) of the protein, such as albumin (pI 5.2–5.9) and immunoglobulin (pI 8–9), which represent approximately 80% of the human blood serum. Such conditions significantly support the values observed in this study. In addition, these results are in agreement with data previously reported by Peng et al. [36], who found that the maximum adsorption of albumin on the surface of the Fe_3O_4 nanoparticle is close to its pI. Therefore, considering that albumin is the most abundant protein in human serum, being the main proteins interfering in the detection of low abundance proteins, and as one of the study objectives is to evaluate methods for the removal of abundant proteins from serum samples, was selected a pH value of 5.5 to accomplish protein depletion using MNPs.

2.4.1.1 Depletion of Abundant Proteins Using MNPs

Since optimum conditions for the treatment of serum samples were found using the nanoparticles, such conditions were applied to obtain the SDS-PAGE profile of the supernatant and to evaluate the removal of high abundance proteins using MNPs. For protein depletion, 25 μL human serum diluted with 75 μL Tris-HCl pH 5.5 were shaken (300 rpm) with 5 μg MNPs (70 nm) for 30 min at 25 °C. After that time, the supernatant and the MNPs containing adsorbed proteins were magnetically separated. The supernatant was evaporated to dryness and reconstituted in appropriate buffer for SDS-PAGE analysis, then the reconstituted proteins were quantified. The quantification results showed the proteins recoveries in the supernatant (about 4.6 μg μL^{-1}) from the serum protein (see bar graph, Fig. 2.8, panel a).

Although the mechanism of protein adsorption on the nanoparticles surface is unclear, many interactions may contribute for the adsorption of proteins in MNPs, such as van der Waals interactions, interactions hydrophobic interactions, and especially electrostatic interactions [20] (Fig. 2.9). In addition, different proteins have different properties, which lead to different binding behaviors in the nanoparticle.

Figure 2.8 (panel b) illustrates the SDS-PAGE gel image, where the supernatant proteins were incubated with MNPs under ideal conditions. When protein depletion using MNPs is compared to crude serum, an increase in band resolution is observed. In addition, the band resolution clearly indicates the depletion of some high molecular weight proteins (>80 kDa), as well as the enrichment of those low molecular weight proteins, when compared to the crude serum. However, this method does not significantly reduce albumin, as highlighted by the intense band in the region close to 66 kDa (Fig. 2.8, panel b). This fact can be explained in terms of protein size,

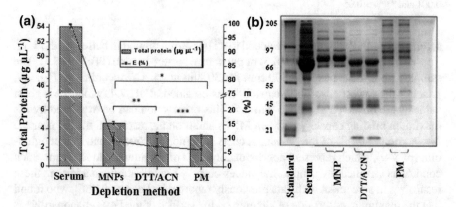

Fig. 2.8 Comparison of depletion methods using MNPs, DTT/ACN and PM. **a** Protein concentration (μg μL^{-1}). Results obtained from three independent experiments. The curve plot represents the protein recovery (%) after depletion. Statistical analysis was performed with t-test and considered significant for $*p < 0.001$, $**p < 0.05$ and $***p > 0.05$. **b** Representative image of SDS-PAGE gel of the blood serum sample after application of the depletion methods, in duplicate. A fixed amount of protein (30 μg) was used for all methods

Fig. 2.9 Representative image of the possible nanoparticle-protein interaction. Several interactions may influence the proteins adsorption on the nanoparticles surface, e.g. electrostatic interaction, generated by the medium pH

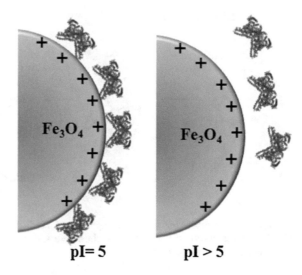

in other words, larger proteins can bind more strongly to MNPs when compared to smaller ones [20, 36]. As albumin is a small protein (~66 kDa), and as there are larger proteins with the pI similar to albumin, competition for adsorption sites in MNPs increase the total removal of albumin from serum.

2.4.1.2 Protein Depletion Using DTT and ACN in Sequential

Recent studies have shown that ACN and DTT can be used as efficient reagents to remove abundant proteins from human blood (serum and plasma) [25, 49]. ACN demonstrated good efficiency in the reduction of high molecular weight proteins from human serum, where DTT proved to be efficient in depleting abundant proteins rich in disulfide bonds, such as albumin and transferrin. Based on these results, both chemical reagents (DTT and ACN) were sequentially combined with the aim of removing high abundance proteins in human blood serum to find differentially abundant proteins with potential for BD biomarker. Thus, human serum was first treated with DTT, and then the supernatant was treated with ACN.

Incubation of 20 μL of serum with 2.2 μL DTT (500 mmol L^{-1}) for 60 min at 37 °C resulted in a white solid precipitate. This fact can be explained by the disruption of the intermolecular and intramolecular disulfide bonds of the proteins (Fig. 2.10). This bonds destruction promotes denaturation of thiol-rich proteins, allowing the agglomeration and subsequent precipitation of such proteins [49].

After precipitation, the supernatant was separated from the precipitate by centrifugation (13,000g). The supernatant was then diluted with water (45 μL) and homogenized using vortex. After drop to drop addition of ACN (85 μL), the solution became turbid and instantaneously observed protein aggregation. This aggregation

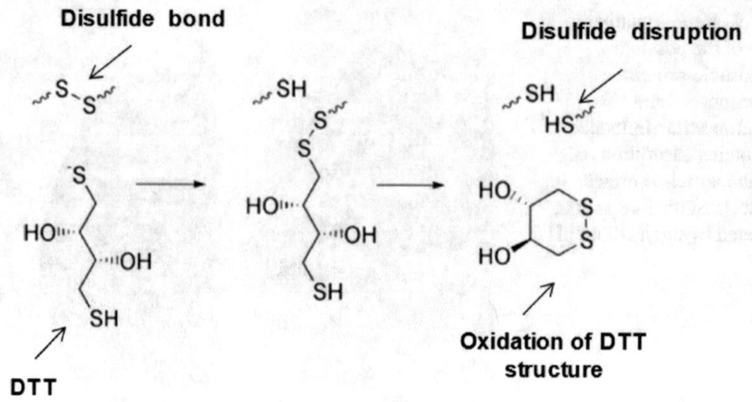

Disulfide bond   **Disulfide disruption**

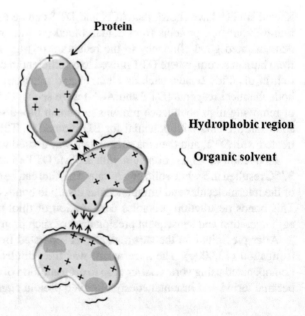

Fig. 2.10 Representative image of the disruption reaction of a disulfide bond made by dithiothreitol (DTT). The disruption of these bonds allows the denaturation of the proteins, and consequently their agglomeration and precipitation

may be caused by the organic solvents reducing the dielectric constant of the protein solution, displacing the water molecules around the hydrophilic regions on the surface of the protein, increasing the electrostatic attraction between the protein molecules (Fig. 2.11). Thus, the proteins aggregate and precipitate. Prior to the electrophoresis analysis, protein quantification was performed. From the quantification, it was found that after depletion, only 3.6 μg μL^{-1} (bar graph, Fig. 2.8, panel a) of the initial concentration was maintained. The SDS-PAGE profile obtained after

Fig. 2.11 Illustration of the precipitation process of proteins using organic solvent. The organic solvent reduces the solution dielectric constant by displacing the water molecules around the hydrophilic regions on the surface of the protein, increasing the electrostatic attractions between the protein molecules of proteins, causing their precipitation

sequential depletion (see Fig. 2.8, panel b) shows a decrease in high molecular weight proteins (characterized by ACN depletion) and also the enrichment of some proteins (characterized by DTT depletion), whose are masked by albumin.

2.4.1.3 Equalization of Proteins Using PM

PM is a commercial product developed by BioRad that allows reducing the dynamic range of proteins in the blood serum. The depletion mechanism is based on the selective adsorption of proteins in a chromatographic medium containing a set of selective peptides. The proteins of low and high abundance are enriched and reduced in a similar way, providing equivalent concentrations, which also allows the detection of those proteins of lower abundance (potential biomarkers). According to the total protein quantification, approximately 3.3 $\mu g \, \mu L^{-1}$ were recovered (bar graph, Fig. 2.8, panel a). Due to the low recovery of the protein mass, new depletions were necessary. About 30 μg were used to perform SDS-PAGE separation (Fig. 2.8, panel b), and it was possible to observe a lower density in the gel bands, confirming a decrease in the dynamic range of proteins in the serum sample.

2.4.1.4 Comparison of the Efficiency of the Depletion Methods

The efficiency and reproducibility of the depletion methods were evaluated by duplicate SDS-PAGE gel analysis. From analyzes, no significant differences were observed between the duplicates of gels (Fig. 2.12), confirming the methods efficiency to simplify the proteome of the blood serum. Not surprisingly, among the evaluated depletion methods, the method that yielded the best results for the removal

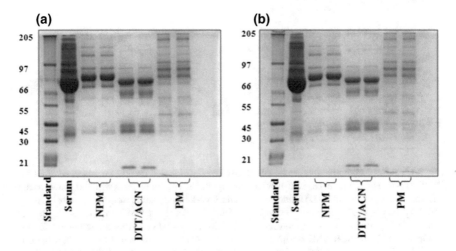

Fig. 2.12 SDS-PAGE analysis. **a** First reply, and **b** second reply

of high abundance proteins was the method using the PM kit. With the PM a lower protein content, $3.3 \pm 0.2 \ \mu g \ \mu L^{-1}$ (n = 3) was observed in the supernatant, while using DTT + ACN resulted in a supernatant with a protein content of $3.6 \pm 0.1 \ \mu g \ \mu L^{-1}$ (n = 3). Depletion using MNPs showed a supernatant with a total protein content of $4.6 \pm 0.2 \ \mu g \ \mu L^{-1}$ (n = 3) (Fig. 2.8, panel a). Although, depletion methods using MNPs and chemical agents (DTT + ACN) show significant removal of abundant proteins as seen in Fig. 2.8, the equalization method using the commercial PM kit significantly decreased the dynamic range of abundant proteins, as well as allowed a enrichment of those smaller proteins abundance (Fig. 2.8, panel b) and therefore was chosen for this study as the best methodology for the treatment of blood serum from different research groups in order to identify differential proteins with potential for BD biomarker.

2.4.1.5 Proteomic Analysis Using 2-D DIGE and NLC-QTOF MS

After removal of the abundant proteins present in patients' blood serum samples (BD, SCZ, OD) and controls (CHF and CHNF) using the PM kit, the 2-D DIGE analyzes were started. For this, approximately 60 µg depleted proteins from each group were labeled with CyDyes dyes. The experimental design described in Table 2.2 was used to carry out labeling of the proteins with Cy3 or Cy5. All gels were used to detect differences between groups. The representative images of the 2-D DIGE gels are shown in Fig. 2.13, panel a. With the computational comparison of 2-D DIGE

(a)   **(b)**

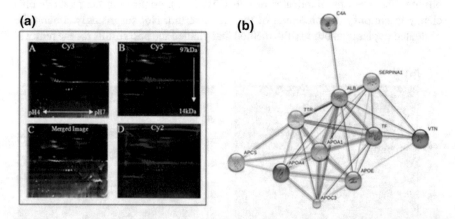

Fig. 2.13 a Representative Images of 2-D DIGE. (A) Cy3-labeled protein fluorescence of HCNF individuals shown as green. (B) Cy5-labeled protein fluorescence of patients with BD shown as red. (C) Mixed image of **a** and **b**. (D) Fluorescence of the Cy2-labeled protein from the mixture of HCNF and BD proteins, used as an internal standard. **b** Global molecular network analysis of the proteins identified in this study. The codes represent: ALB—albumin; APOA1—Apolipoprotein A-I; C4A—Complement C4-A; SERPINE 1-Alpha-1-antitrypsin; VTN—Vitronectin; APOC3—Apolipoprotein C-III; TTR—Transteritine; APOA4—Apolipoprotein A-IV; TF—Transferrin; SAMP—Amyloid Serum P

gel images, 37 protein spots were found to be differentially abundant ($p < 0.05$, t-Student). These spots exhibited variation ≥ 2 times the variation of the mean value of the intensity of standardized spots in the serum of SCZ, BD and OD patients when compared to controls (CHF and CHNF).

Of these 37 spots differentially found, 13 different proteins were identified. The list of identified proteins is given in Table 2.3, which includes MS analysis information (combined peptides, coverage and score) as well as information from the 2-D DIGE (regulatory factor, p-values) analysis.

When comparing the HCNF group and the HCF group, five spots were found to be differentially abundant. From these spots, only three proteins (transthyretin, TTR, kappa immunoglobulin, IgK, apolipoprotein A-IV, ApoA4) were identified. TTR and IgK were found with lower abundance, while ApoA4 was found to be highly abundant in the CHNF sample when compared to CHF. For comparative analysis between CHNF and BD groups, 23 abundantly differentiated spots were found, however, only six different proteins were identified. The low number of proteins identified is justified in terms of isoforms (see Table 2.3). Proteins (albumin, Alb, and apolipoprotein AI, ApoA1) were detected with higher abundance and four proteins with lower abundance (complement C4-A, C4A, alpha-1-antitrypsin, SerpinA1 and apolipoprotein E, ApoE). the BD group is compared with CHNF.

In the comparative study involving BD and OD groups, four proteins (Alb; C4A, vitronectin, VTN and apolipoprotein C-III, ApoC3) were identified and five spots were detected as differentially abundant. All identified proteins were found to be less abundant in OD when compared to BD. In comparative analyzes involving CHF versus group BD, eight spots were detected as differentially abundant, and seven proteins (transferrin, TR and six ApoA1 isoforms) were identified. Only TTR was found with lower abundance, while all ApoA1 isoforms were detected higher in BD compared to HCF. Finally, when comparing the SCZ and BD groups, six spots were found to be differentially abundant, and only three proteins (C4A, complement component C4-B, C4B and amyloid serum P, Samp) were identified. All proteins were found to have lower serum abundance for BD compared to SCZ. Figure 2.14 shows the representative gel profile, highlighting the individual spots with a 3D view of these thirteen different proteins, differentially abundant in the BD serum.

2.4.1.6 Proof of Concept: Understanding the Relation Between Identified Proteins and the BD

To verify the interactions and functions of the proteins identified in this study with the BD, a network of molecular interactions was constructed. Network construction was done using the STRING v10 software (www.string-dp.org). Thirteen proteins that showed significant changes in the experimental groups (TAB, ESQ, OTM) and healthy control groups (HCF and HCNF) were included in the analysis. Each identified protein was converted into its gene and mapped to its corresponding gene object in the STRING knowledge base. A network was generated and the direct or indirect relationships exhibited by proteins within the network are shown in Fig. 2.13, panel b.

Table 2.3 Information of the proteins identified in the proteomic study

Spot #	CHNF versus BD Protein	Score	pI a	pI b	MW (Da) a	MW (Da) b	Pept. match	Cov. (%)	Abundant factor	p-values
1	Albumin	202	5.92	5.90	71,317	128,753	13	21	+5.85	0.004
2	Albumin	910	5.92	5.99	71,317	131,773	43	56	+5.45	0.051
3	Albumin	790	5.92	6.08	71,317	131,773	43	57	+5.2	0.052
4	–	–	–	–	–	–	–	–	–	–
5	–	–	–	–	–	–	–	–	–	–
6	Apolipoprotein A-I	33	5.56	6.04	30,759	68,806	6	21	+2.48	0.050
7	Complement C4-A	37	6.65	6.11	19,4247	70,148	3	1	−2.56	0.022
8	Complement C4-A	140	6.65	6.53	19,4247	68,365	8	4	−2.33	0.023
9	Complement C4-A	71	6.65	6.76	19,4247	69,251	9	5	−2.99	0.021
10	Alpha-1-antitrypsin	66	5.37	4.87	46,878	43,544	10	22	- 2.79	0.004
11	Apolipoprotein A-I	148	5.56	5.05	30,759	39,024	18	49	+2.70	0.007
12	Apolipoprotein A-I	42	5.56	4.37	30,759	34,972	5	15	+3.76	0.057
13	Apolipoprotein E	128	5.65	6.48	36,246	44,509	9	34	−2.40	0.055
Spot #	BD versus OD Protein	Score	pI a	pI b	MW (Da) a	MW (Da) b	Pept. match	Cov. (%)	Abundant factor	p-values
14	–	–	–	–	–	–	–	–	–	–
15	Vitronectin	80	5.55	5.44	55,069	67,058	8	14	−2.13	0.022
16	Albumin	436	5.92	5.56	71,317	67,926	26	36	−3.16	0.055
17	Complement C4-A	31	6.65	4.15	194,247	21,726	2	1	−2.65	0.050
18	Apolipoprotein C-III	52	5.23	4.30	10,845	11,324	2	27	−2.65	0.027

(continued)

Table 2.3 (continued)

Spot #	CHF versus CHNF	Score	pI		MW (Da)		Pept match	Cov. (%)	Abundant factor	p-values
	Protein		a	b	a	b				
19	–	–	–	–	–	–	–	–	–	–
20	Transthyretin	82	5.52	4.99	15,991	9516	2	8	−2.39	0.026
21	Ig kappa chain C region	77	5.58	5.55	11,773	9205	5	51	−3.52	0.057
22	–	–	–	–	–	–	–	–	–	–
23	Apolipoprotein A-IV	44	5.28	4.14	45,371	36,141	8	20	+4.00	0.003

Spot #	CHF versus BD	Score	pI		MW (Da)		Pept. match	Cov. (%)	Abundant factor	p-values
	Protein		a	b	a	b				
24	Apolipoprotein A-I	41	5.56	4.89	30,759	36,540	6	19	+2.83	0.050
25	Apolipoprotein A-I	369	5.56	5.09	30,759	36,141	27	63	+3.07	0.051
26	Apolipoprotein A-I	292	5.56	5.01	30,759	35,358	26	58	+4.16	0.058
27	Apolipoprotein A-I	228	5.56	4.95	30,759	35,748	24	56	+3.09	0.057
28	Apolipoprotein A-I	326	5.56	4.89	30,759	35,748	23	58	+4.12	0.041
29	Apolipoprotein A-I	193	5.56	4.85	30,759	36,540	16	51	+4.34	0.054
30	Transferrin	51	6.81	5.03	79,280	93,109	7	13	−2.79	0.004
31	–	–	–	–	–	–	–	–	–	–

(continued)

Table 2.3 (continued)

Spot #	SCZ versus BD	Score	pI		MW (Da)		Pep. match	Cov (%)	Abundant factor	p-values
	Protein		a	b	a	b				
32	Complement C4-A	116	6.65	6.53	19,4247	68,365	6	2	−2.33	0.023
	Complement C4-B	116	6.73	6.53	194,212	68,365	6	2	- 2.33	0.023
33	Complement C4-A	320	6.65	6.53	194,247	68,365	23	12	−2.56	0.022
	Complement C4-B	320	6.73	6.53	194,212	68,365	23	12	−2.56	0.022
34	Complement C4-A	60	6.65	6.71	194,247	147,965	5	2	−3.58	0.008
	Complement C4-B	60	6.73	6.71	194,212	147,965	5	2	−3.58	0.008
35	Complement C4-A	153	6.65	6.89	194,247	127,270	9	6	−2.98	0.016
	Complement C4-B	153	6.73	6.89	194,212	127,270	9	6	−2.98	0.016
36	Complement C4-A	156	6.65	6.97	194,247	122,921	3	2	−5.27	0.020
	Complement C4-B	156	6.73	6.97	194,212	122,921	3	2	−5.27	0.020
37	Serum amyloid P-component	157	6.10	7.02	25,485	21,280	8	33	−2.00	0.021

pI isoelectric point; *MW* molecular weight; *pep* peptides; *cov* coverage
[a] teoric value; [b] experimental values

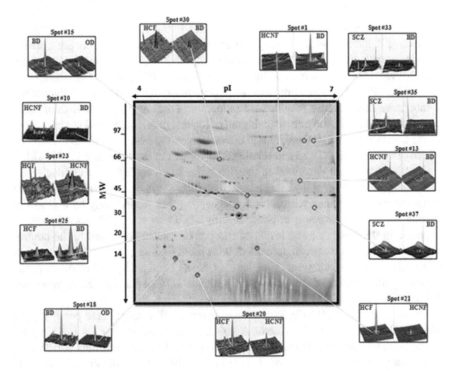

Fig. 2.14 3D view of the individual spots of the thirteen proteins identified as differentially abundant in the blood serum of HCNF, CHF, BD, SCZ and OD using the Decyder program. The spots are: Spot #1-albumin; spot #10-alpha-1-antitrypsin; spot #13-apolipoprotein E; spot #15-vitronectin; spot #18-apolipoprotein C-III; spot #20-transthyretin; spot #21-Ig kappa; spot #23-Apolipoprotein A-IV; spot #25-apolipoprotein A-I; spot #30-transferrin; spot #33-Complement C4-A; spot #35-Complement C4-B; spot #37-Amyloid Serum P

The network showed 11 proteins related to the biological processes of the brain. Two proteins (C4B and IgK) were not recognized by the STRING software and therefore are not presented in the network analysis. All identified proteins were functionally categorized based on Gene Ontology (GO) annotations using the Universal Protein Resource (Uniprot) program (www.uniprot.org). The major categories of functions associated with the network included molecular transporter, inflammatory interactions and immune response, enzyme regulator and cell signaling.

Based on the global molecular interactions network (Fig. 2.13, panel b), proteins such as ApoA1, ApoE, ApoC3, ApoA4, Samp, SerpinA1, TTR, IgK, Alb, VTN, TR, C4A and C4B are associated with the inflammatory response. These proteins are also known as acute phase proteins (AP), and are directly or indirectly related to proinflammatory cytokines [13, 18]. These cytokines are excreted by inflammatory cells in response to some injury, such as neuroinflammation [13]. Although the neurochemical mechanism of BD pathophysiology is still not fully understood, an association between bipolar disorder and immunological and inflammatory functioning may be a mechanism to explain the pathophysiology of BD [13, 18]. Thus, inflammatory proteins may play important roles in the cognitive decline observed in patients diagnosed with BD [13]. Inflammation in neurotransmitters has a strong impact on the patient's metal health. Therefore, the presence of a proinflammatory state activates the enzyme indolamine 2,3-dioxygenase (IDO), which degrading tryptophan and serotonin. The stimulation of IDO increases the production of tryptophan catabolites. This stimulus may cause a reduction in the energy metabolism of mitochondria. In addition, the free radicals generation and peroxide, as well as an increase in the neuroexcitatory and neurotoxic effect can lead to neuropsychiatric disorders. These effects are modulated and regulated by inflammatory cytokines [7, 13, 18].

Similar results are described in the literature. In an earlier study conducted by our group, Sussulini et al. [44, 45] reported an increase in the level of ApoA1 in the serum of patients treated with lithium, suggesting that ApoA1 may be a potential marker for lithium response in patients with BD. Dean et al. [10] also reported a decrease in the ApoE plasma levels of BD patients when compared to the control and schizophrenic groups. In agreement with these findings, the results obtained in this study showed an increase of the level of ApoA1 and a decrease of ApoE in the serum of patients with BD in comparison with the OD group and the SCZ group (see Table 2.3).

In addition, this study also demonstrates levels significantly altered positively for the complement proteins C4 (factor A and B), immunoglobulins (as IgK), alpha-antitrypsin, albumin and transferrin in patients with BD. These results corroborate with those reported by Giusti et al. [17] and Herbert et al. [18] that showed differences in the expression of these proteins in peripheral blood mononuclear cells and peripheral lymphocytes, suggesting the relationship between BD, acute phase proteins, and inflammatory response.

2.5 Partial Conclusion

From the proteomic study, we conclude that the main objectives of this work were widely achieved. Among the depletion methods evaluated, the PM kit presented the best strategy to remove proteins of high abundance and was therefore chosen for the treatment of serum BD, SCZ, OD, HCNF and HCF. However, although depletion methods using MNPs and chemical agents (DTT + ACN) have not been chosen, it does not mean that they were inefficient for removal of the high abundance proteins present in the serum of blood, on the contrary, these methods have been shown to be efficient. However, the PM kit showed better resolution of the protein bands separation when compared to the other methods of depletion and therefore it was applied in the comparative study of the BD with the other study groups. By comparing 2-D DIGE gel images, 37 protein spots were found to be differentially abundant ($p < 0.05$, Student's t-test). These spots had an average variation greater than 2 times the variation of the mean value of the intensity of standard spots in the serum of SCZ, BD and OD in comparison with the controls (HCF and HCNF). From these detected spots, 13 different proteins were identified: ApoA1, ApoE, ApoC3, ApoA4, Samp,

SerpinA1, TTR, IgK, Alb, VTN, TR, C4A and C4B. Through the network of global interaction, these proteins, known as acute phase proteins, work together and are directly or indirectly related to pro-inflammatory cytokines, which act in response to some inflammatory lesion, such as neuroinflammation, justifying, therefore, the changes found in the blood serum sample of patients with BD when compared with the controls.

References

1. Araújo JE, Lodeiro C, Capelo JL, Rodriguez-Gonzàlez B, Santos A, Santos HM, Fernández-Lodeiro J (2015) Novel nanocomposites based on a strawberry-like gold-coated magnetite (Fe@Au) for protein separation in multiple myeloma serum samples. Nano Res 8(4):1189–1198
2. Atkinson and Biomarkers Definitions Working Group (2001) Biomarkers and surrogate endpoints: preferred definitions and conceptual framework. Clin Pharmacol Ther 69(3):89–95
3. Chen J, Wang Y, Ding X, Huang Y, Xu K (2014) Analytical methods on hydroxy functional ionic liquid-modified magnetic nanoparticles. Anal Methods 6:8358–8367
4. Chudal R, Sourander A, Polo-Kantola P, Hinkka-Yli-Salomaki S, Lehti V, Sucksdorff D, Gissler M, Brown AS (2014) Perinatal factors and the risk of bipolar disorder in finland. J Affect Disord 155:75–80
5. Corrigan L, Browne J, Fitzgerald K, Chan GC, Kennely S, Marry J, Nash M, Ryan D, Ryan D, Fallon N, Steen G, Casey M, Walsh JB, Lee TC, Daly J (2011) 2D Dige as a strategy for identifying differentially expressed proteins in postmenopausal women with osteopenia and osteoporosis. Bone 48:194
6. de Jesus JR, Campos BK, Galazzi RM, Martinez JL, Arruda MA (2015) Bipolar disorder : recent advances and future trends in bioanalytical developments for biomarker discovery. Anal Bioanal Chem 407:661–667
7. de Jesus JR, Pessôa GS, Sussulini A, Martinez JLC, Arruda MAZ (2016) Proteomics strategies for bipolar disorder evaluation: from sample preparation to validation. J Proteom 145:187–196
8. de Jesus JR, Galazzi RM, Lima TB, Banzato CEM, Almeida Lima e Silva LF, Rosalmeida Dantas C, Gozzo FC, Arruda MAZ (2017a) Simplifying the human serum proteome for discriminating patients with bipolar disorder of other psychiatry conditions. Clin Biochem 50(18):1118–1125
9. de Jesus JR, Silva Fernades R, Souza Pessôa G, Raimundo IM Jr, Arruda MAZ (2017b) Depleting high-abundant and enriching low-abundant proteins in human serum: an evaluation of sample preparation methods using magnetic nanoparticle, chemical depletion and immunoaffinity techniques. Talanta 170:199–209
10. Dean B, Digney A, Sundram S, Thomas E, Scarr E (2008) Plasma apolipoprotein E is decreased in schizophrenia spectrum and bipolar disorder. Psychiatry Res 158(1):75–78
11. Deng J, Yang Y, Wang X, Tiangang L (2014) Strategies for coupling solid-phase microextraction with mass spectrometry. Trac Trends in Analytical Chemistry 55:55–67
12. Domon B, Aebersold R (2006) Mass spectrometry and protein analysis. Science 312(5771):212–217
13. Fan NJ, Kang R, Ge XY, Li M, Liu Y, Chen HM, Gao CF (2014) Identification alpha-2-hs-glycoprotein precursor and tubulin beta chain as serology diagnosis biomarker of colorectal cancer. Diagn Pathol 9(53):1–11
14. Fernández-Costa C, Reboiro-Jato M, Fdez-Riverola F, Ruiz-Romero C, Blanco FJ, Martínez JLC (2014) Sequential depletion coupled to C18 sequential extraction as a rapid tool for human serum multiple profiling. Talanta 125:189–195
15. Fulford D, Peckham AD, Johnson K, Johnson S (2014) Emotion perception and quality of life in bipolar disorder. J Affect Disord 152–154:491–497

16. Geoffroy PA, Boudebesse C, Bellivier F, Lajnef M, Henry C, Leboyer M, Scott J, Etain B (2014) Sleep in remitted bipolar disorder: a naturalistic case-control study using actigraphy. J Affect Disord 158:1–7
17. Giusti L, Mantua V, Da Valle Y, Ciregia F, Ventroni T, Orsolini G, Donadio E, Giannaccini G, Mauri M, Cassano GB, Lucacchini A (2014) Search for peripheral biomarkers in patients affected by acutely psychotic bipolar disorder: a proteomic approach. Mol BioSyst 10(6):1246–1254
18. Herberth M, Koethe D, Levin Y, Schwarz E, Krzyszton ND, Schoeffmann S, Ruh H, Rahmoune H, Kranaster L, Schoenborn T, Leweke MF, Guest PC, Bahn S (2011) Peripheral profiling analysis for bipolar disorder reveals markers associated with reduced cell survival. Proteomics 11(1):94–105
19. Hilty DM, Leamon MH, Lim RF, Kelly RH, Hales RE (2006) A review of bipolar disorder in adults. Psychiatry 3(9):43–55
20. Hu Z, Zhang H, Zhang Y, Wu R, Zou H (2014) Nanoparticle size matters in the formation of plasma protein coronas on Fe_3O_4 nanoparticles. Colloids Surf B 121:354–361
21. Huang JTJ, Leweke FM, Oxley D, Wang L, Harris N, Koethe D, Gerth CW, Nolden BM, Gross S, Schreiber D, Reed B, Bahn S (2006) Disease biomarkers in cerebrospinal fluid of patients with first-onset psychosis. Plos Med 3(1.1):2145–2158
22. Iavarone F, Mellis M, Platania G, Cabras T, Manconi B, Petruzzelli R, Cordaro M, Siracusano A, Faa G, Messana I, Zanasi M, Castagnola M (2014) Characterization of salivary proteins of schizophrenic and bipolar disorder patients by top-down proteomics. J Proteom 103:15–22
23. Javanmard M, Emaminejad S, Gupta C, Provine J, Davis RW, Howe RT (2014) Sensors and actuators B: chemical depletion of cells and abundant proteins from biological samples by enhanced dielectrophoresis. Sens Actuators B Chem 193:918–924
24. Jessen F, Wulff T (2015) Triton x-114 cloud point extraction to subfractionate blood plasma proteins for two-dimensional gel electrophoresis. Anal Biochem 485:1–7
25. Kay R, Barton C, Ratcliffe L, Matharoo-Ball B, Brown P, Roberts J (2008) Enrichment of low molecular weight serum proteins using acetonitrile precipitation for mass spectrometry based proteomic analysis. Rapid Commun Mass Spectrom 22:3255–3260
26. Kinter M, Sherman NE (2000) Protein sequencing and identification using tandem mass spectrometry. Wiley, New York, pp 29–63. https://doi.org/10.1002/0471721980
27. Kurdal E, Tanriverdi D, Savas HA (2014) The effect of psychoeducation on the functioning level of patients with bipolar disorder. West J Nurs Res 36(3):312–328
28. Leboyer M, Soreca I, Scott J, Frye M, Henry C, Tamouza R, Kupfer DJ (2012) Can bipolar disorder be viewed as a multi-system inflammatory disease? J Affect Disorders 141(1):1–10
29. Li F, Seillier-Moiseiwitsch F, Korostyshevskiy VR (2011) Region-based statistical analysis of 2d page images. Comput Stat Data Anal 55(11):3059–3072
30. Lu AH, Salabas EL, Schüth F (2007) Magnetic nanoparticles: synthesis, protection, functionalization, and application. Angewandte Chemie - International Edition 46(8):1222–1244
31. Mantini D, Petrucci F, Del Boccio P, Pieragostino D, Di Nicola M, Lugaresi A, Federici G, Sacchetta P, Di Llio C, Urbani A (2008) Independent component analysis for the extraction of reliable protein signal profiles from maldi-tof mass spectra. Bioinformatics 24(1):63–70
32. Marmol F (2008) Lithium: bipolar disorder and neurodegenerative diseases possible cellular mechanisms of the therapeutic effects of lithium. Prog Neuropsychopharmacol Biol Psychiatry 32(8):1761–1771
33. Matsumoto J, Sugiura Y, Yuki D, Hayasaka T, Goto-Inoue N, Zaima N, Kunii Y, Wada A, Yang Q, Nishiura K, Hashizume Y, Yamamoto T, Ikemoto K, Setou M, Niwa S (2011) Abnormal phospholipids distribution in the prefrontal cortex from a patient with schizophrenia revealed by matrix-assisted laser desorption/ionization imaging mass spectrometry. Anal Bioanal Chem 400:1933–1943
34. Moraes MCB, Lago CL (2003) Espectrometria de massas com ionização por "electrospray" aplicada ao estudo de espécies inorgânicas e organometálicas. Quim Nova 26(4):556–563
35. Patel S (2012) Role of proteomics in biomarker discovery and psychiatric disorders: current status, potentials, limitations and future challenges. Expert Rev Proteom 9:249–265

36. Peng ZG, Hidajat K, Uddin MS (2004) Adsorption of bovine serum albumin on nanosized magnetic particles. J Colloid Interface Sci 271(2):277–283
37. Rabilloud T, Lelong C (2011) Two-dimensional gel electrophoresis in proteomics: a tutorial. J Proteom 74(10):1829–1841
38. Redpath HL, Cooper D, Lawrie SM (2013) Imaging symptoms and syndromes: similarities and differences between schizophrenia and bipolar disorder. Biol Psychiat 73(6):495–496
39. Renner J, Almeida C, Phillips ML (2012) Distinguishing between unipolar depression and bipolar depression: current and future clinical and neuroimaging perspectives. Biol Psychiat 73(2):111–118
40. Rico Santana N, Muniz EZ, Cocho D, Bravo Y, Mederos RD, Marti-Fábregas J (2014) Analysis of peptidome profiling of serum from patients with early onset symptoms of ischemic stroke. J Stroke Cerebrovasc Dis 23(2):235–240
41. Roche S, Tiers L, Provansai M, Seveno M, Piva MT, Jouin P, Lehmann S (2009) Depletion of one, six, twelve or twenty major blood proteins before proteomic analysis: the more the better? J Proteom 72(6):945–951
42. Sokolowska I, Ngounou AG (2013) The potential of biomarkers in psychiatry: focus on proteomics. J Neural Transm 122:1–9
43. Sussulini A, Prando A, Maretto DA, Poppi RJ, Tasic L, Banzato CEM, Arruda MAZ (2009) Metabolic profiling of human blood serum from H NMR spectroscopy and chemometrics. Anal Chem 81(23):9755–9763
44. Sussulini A, Dihazi H, Banzato CE, Arruda MA, Stuhmer W, Ehrenreich H, Jahn O, Kratzin HD (2011a) Apolipoprotein AI as a candidate serum marker for the response to lithium treatment in bipolar disorder. Proteomics 11(2):261–269
45. Sussulini A, Banzato CEM, Arruda MAZ (2011) Exploratory analysis of the serum ionomic profile for bipolar disorder and lithium treatment. Int J Mass Spectrom 307(1–3):182–184
46. Timerbaev A, Pawlak K, Gabbiani C, Messori L (2011) Recent progress in the application of analytical techniques to anticancer metallodrug proteomics. TrAC Trends Anal Chem 30(7):1120–1138
47. Tonge R, Shaw J, Middleton B, Rowlinson R, Rayner S, Young J, Pognan F, Hawkins E, Davison M (2001) Validation and development of fluorescence two-dimensional differential gel electrophoresis proteomics technology. Proteomics 1(3):377–396
48. Von thun und hohenstein-blaul N, Funke S, Grus FH (2013) Tears as a source of biomarkers for ocular and systemic diseases. Experim Eye Res 117:126–137
49. Warder SE, Tucker LA, Strelitzer TJ, Mckeegan EM, Meuth JL, Jung PM, Saraf A, Singh B, Lai-Zhang J, Gagne G, Rogers JC (2009) Reducing agent-mediated precipitation of high-abundance plasma proteins. Anal Biochem 387(2):184–193
50. Yang ST, Liu Y, Wang YW, Cao A (2013) Biosafety and bioapplication of nanomaterials by designing protein-nanoparticle interactions. Small 9(9–10):1635–1653

Chapter 3
Application of the Ionomic Strategy to Evaluate Difference in Metal Ion Concentration Between Patients with Bipolar Disorder and Other Psychiatric Disorder

3.1 Objectives

3.1.1 General Objective

Design the ionomic profile of the blood serum of individuals with BD, in order to identify differences in the concentration of metal ions, allowing differentiating patients diagnosed with BD from other psychiatric disorders, such as schizophrenia, as well as from healthy individuals.

3.1.2 Specific Objectives

- Evaluate differences in Zn, Cu, Fe, Li, Cd and Pb concentrations in serum samples from individuals with bipolar disorder, schizophrenia and healthy individuals;
- Propose a new methodology based on ultrasonic energy that allows to extract the metals of interest from the human blood serum, in a short period of time;
- Evaluate the best composition of the extractive solution;
- Evaluate the best condition of ultrasound potency for the extraction of the analytes;
- Evaluate the optimal sonication time applied in the extraction process;
- Compare the efficiency of the proposed method with the microwave assisted decomposition (main methodology applied in this type of study);
- Apply the method proposed in blood serum samples from healthy individuals and patients diagnosed with bipolar affective disorder and schizophrenia, in order to determine possible differences in the concentrations of metal ions in samples of each study subjects;
- Establish possible relationships between the differences found in metal ion concentration and the pathophysiology of bipolar disorder.

© Springer Nature Switzerland AG 2019

J. R. de Jesus, *Proteomic and Ionomic Study for Identification of Biomarkers in Biological Fluid Samples of Patients with Psychiatric Disorders and Healthy Individuals*, Springer Theses, https://doi.org/10.1007/978-3-030-29473-1_3

3.2   Review of the Literature

Overview presented below will address the main issue related to this chapter.

3.2.1   General Aspects of the Functions of the Chemical Elements in the Biological System

Some chemical elements in their ionic forms, free or associated with other chemical species, are present in living organisms and play fundamental roles in the maintenance and biological system functioning [31]. Many of these essential elements, such as Zn, Cu and Fe, act as cofactors for enzymes that participate in various biological events, such as oxygen transport, inhibition of free radical formation, structural organization of macromolecules, hormonal activity, and others [16]. However, since there are elements that are fundamental for the life maintenance, there are also those elements that when in contact with the biological system of a living organism, can cause diverse toxicological reactions, promoting biochemical changes in the biological system, being able to cause the death immediately (when exposure to high concentrations) or slowly (when there is exposure and accumulation of low concentrations). These elements are used in several industry fields and are considered as one of the main environmental contaminants, causing death by exposure [11]. In addition, many essential elements can also be toxic, since the toxicity depends on the concentration and the chemical species of the element present in the organism.

In view of this, it is necessary to study the relationship of chemical elements and the triggering of important human diseases. In this sense, the metallomics emerges as an important evaluation strategy.

3.2.2   Metallomics

Metallomics is the science field that integrate the research fields related to biometals. Metallomics is directly related to genomics and proteomics, since the synthesis and metabolic functions of genes and proteins generally occur in the presence of some metal ions or metalloenzymes, acting as biological catalysts in the regulation of biological reactions and the physiological functions of an organism [26].

Metallomic approach is based on the metallome study which represents the whole set of all metallic and semi-metallic species present in a biological system, taking into account their identity and concentration [32].

Metallomic study can provide important information about certain organisms, such as: (i) distribution profile within cells in specific tissues; (ii) the coordination environment; and (iii) the concentration of the individual metal species [29].

Metalloproteins and proteins bound to metal or semi-metal are responsible for many metabolic processes, such as: conversion of biological energy into photosynthesis and respiration, or markers of processes that govern gene expression and regulation [10]. However, despite their vital role as regular cofactors, metals may also be highly toxic and be functionally associated with many diseases [19].

Through metallomics, it is possible to explore elements that regulate metabolic reactions that can act as etiological agents of neurological disorders, such as BD and schizophrenia [28, 29, 32]. Metallomics can contribute to design the ionomic profiles (study of free ions into biological system) and metalloproteomics (study of metals or semimetals bound to proteins) in control subjects, compared with patients with different diseases, such as those mentioned above. The evaluation of the metallomic profiles of these individuals allows identifying possible specific markers, and thus collaborating in the diagnosis and therapeutic treatment of these pathologies.

The analytical strategy involving metallomic studies includes two important main components: (i) sample preparation step, which allows obtaining the analytes free interferers; and (ii) is a sensitive detector for the elements quantification.

In terms of metals quantification in biological fluids well-founded analytical methodologies are required [12]. Typically, metal analysis is performed by techniques of significant sensitivity, such as, atomic fluorescence spectrometry (AFS), X-ray fluorescence (XRF), atomic absorption spectrometry (AAS), inductively coupled plasma optical emission spectrometry (ICP-OES) and inductively coupled plasma mass spectrometry (ICP-MS). In recent years, ICP-MS has been the technique of choice for the determination of different elements in different types of samples at concentrations at the level of ng L^{-1} to μg L^{-1} [2, 5, 12].

As ICP-MS was the technique applied in this study, an overview of this technique will be discussed below.

3.2.3 ICP-MS

The versatility of ICP-MS (Fig. 3.1) in terms of sensitivity and selectivity, makes it an appropriate technique for analysis of multielement determination in different samples [5]. Among the main advantages of this technique are the wide linear dynamic range, the low limit of detection, the high speed of analysis, and the ability to obtain simple spectra and the ability for isotopic analysis [4, 5, 12]. An example of the ICP-MS importance in clinical analysis of metals refers to the determination of Pb in the blood. Currently, the determination of Pb in the blood is only possible using ICP-MS, due to the low concentrations found, generally values <1 μg L^{-1} [4, 18].

In this technique, the sample is introduced in solution form. Sample introducer system is characterized for presents a nebulizer (pneumatic or ultrasonic), which forms an aerosol; a nebulizer camera that separating the larger drop from the smaller ones coming from the nebulizer. The aerosol with the analyte is then carried into the plasma by the flow of the nebulizer gas, where the analyte suffers desolvation,

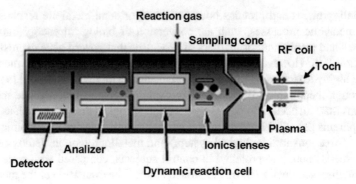

Fig. 3.1 Representative image of an ICP-MS. The figure shows the various components present in such apparatus for the determination of different metals in different samples

vaporization, atomization and ionization. Formation of oxides can also occur due to temperature drop by the gas outlet of the central channel [30].

The ICP source produces ions for the mass spectrometer (MS). To transport the ions formed to the MS, a multiple stage vacuum pump interface is used. The ions formed by the plasma are brought to the MS at low pressure by means of a sampling cone. After passing through a second cone (skimmer cone), the ions are focused on the mass analyzer using a series of lenses (electrodes with different voltages), the ions are then separated into the mass analyzer (quadrupole) according to their mass/charge (m/z) ratio [4, 30].

Plasma is formed in a quartz torch, which consists of three concentric tubes through which argon gas (Ar) flows. External flow is known as central gas or "plasma" gas, it is responsible for maintaining ICP. The carrier gas flow is called "auxiliary", this gas is used to support the plasma and keep it away from the sides of the quartz torch. The internal gas flow is known as nebulizer gas and transports the analyte solution to the plasma. Plasma is formed when a spark, originating from the Tesla coil, is used to produce electrons that are accelerated by the electric and magnetic fields having energy to ionize the argon gas. The gas remains electrically neutral, and this makes it a good conductor of electricity, so part of the plasma energy is transferred to excite and ionize the analyte [30].

Although ICP-MS is a powerful technique in the determination of heavy metals, one of its major problems is its susceptibility to interference [13]. These interferences can be divided between non-spectral and spectral interferences [2, 13].

Non-spectral or physical interferences are related to the properties of the samples, such as viscosity and volatility [2]. Such interference may interfere with the method of introduction and ionization of the sample in plasma. There are several methods applied to detect and eliminate this type of interference, such as the use of internal standard free of spectral interferences (indium and rhodium), addition of standard, isotope dilution, matrix adjustment and others [2, 4].

The spectral interferences present in the ICP-MS technique can be divided into isobaric interferences and polyatomic interferences:

- Isobaric interference occurs when an isotope of one element overlaps the reading of an isotope of another element with the same nominal mass [13].
- Polyatomic interferences are formed from ionized species that are produced in regions of low temperature of the plasma or interface region between the plasma and the mass filters. These polyatomic ions may interfere with the detection of some isotopes of the same nominal mass and are produced from argon gas and other gases such as oxygen and nitrogen as well as reagents used in the preparation of samples such as sulfuric (H_2SO_4), hydrochloric (HCl) and hydrofluoric (HF), or even the matrix itself (salts and ions) [30, 32].

Besides these interferers, there are also the oxides, which are species that can be derived from the sample and/or the recombination of the ions in the plasma [4, 32]. Increasing the plasma temperature, decreasing nebulizer gas flow, introducing a cooled nebulizer chamber (4 °C) or an ultrasonic nebulizer may to eliminate the interferers (Thomas et al. 2008).

Another commonly used option to minimize or even eliminate interference in ICP-MS is the use of a reaction and collision cell (DRC). DRC-ICP-MS is a versatile practice that allows the elimination of interferers, due to the use of gases (CH_4, O_2, NH_3, He and others) that react or collide with the interfering ions, allowing accurate analytes detection.

As an example of the application of ICP-MS to determine the influence of metals ions on important human diseases, we mention the work of Ahmed and Santosh [3]. In this study, the authors evaluated the variation in the concentration of 31 elements in patients with Parkinson's disease and healthy individuals using ICP-MS. As a conclusion, the authors showed that Al, Cu, Fe, Mn, and Zn showed significant differences in concentrations.

Another work that shows the ICP-MS importance in study of human diseases was carried out by Sussulini et al. [28]. In this study, the researchers promoted an exploratory analysis of the ionomic profile in blood serum of patients suffering from BD in order to evaluate the influence of the drug (lithium carbonate) on the biological system during treatment. For this, patients diagnosed with the disease under different treatments (BD under treatment with lithium (n = 15) and BD under treatment with other drugs (n = 10)) were evaluated. As conclusion, the researchers found 15 metal ions (As, B, Cl, Cr, Fe, K, Li, Mg, P, S, Se, Si, Sr and Zn) with significantly altered concentrations between the evaluated groups, to the drug (lithium carbonate) used in the treatment of patients with BD.

3.2.4 Sample Preparation for Metals Determination in Human Blood

For metals determination in blood, different methods for sample preparation are employed and reported in the literature, such as simple dilution [9, 21], or wet digestion [3] or microwave assisted digestion [6]. Heitland and Koster [11], for example, reported a study that evaluated 37 elements in 130 human blood samples from workers in the risk area. For the analysis, the authors used a 1/10 (v/v) dilution of a solution containing 0.1% (v/v) Triton-X and 0.5% (v/v) ammonia as sample preparation and ICP-MS as analysis technique. According to the authors, this study allows physicians and biochemists to understand the metals behavior in biological system of workers exposed to the metals studied.

Ahmed and Santosh [3] proposed to evaluate the influence of 31 elements in blood of patients with Parkinson's disease and controls by ICP-MS, using wet decomposition with 0.5 mL concentrated nitric acid (HNO_3) as sample preparation. With this methodology, the researchers found significant differences in the concentration of five elements (Al, Cu, Fe, Mn, and Zn) between patients and controls.

However, although the simple dilution of biological samples using specific detergents or wet decomposition presents expressive results for the analysis of determination of different metals in biological fluids, as well as allowing a certain practicality in the sample preparation, mainly for the analysis of a relative to sample quantities, such procedures have some limitations, such as (i) loss of instruments sensitivity, since simple dilution using salts and detergent may cause nebulizer clogging and sampling cone due to the accumulation of organic components from the matrix and detergent [4]; (ii) sample contamination risk; (iii) and risk of loss analytes volatile.

Thus, in general, assisted microwave digestion is the main strategy of sample preparation applied in metallomics studies due to its high efficiency and robustness, in addition to avoid the loss of volatile analytes and allowing a decomposition free of contamination [24, 33]. However, while microwave ovens have several advantages such as those cited above, the main limitation of these procedures is the high time available for cooling the reactors prior to their opening. In this sense, the metals extractions using high power ultrasound have been highlighted to obtain elementary quantitative recoveries of several matrices (food and environmental). Extraction using ultrasound, especially with cup-horn sonoreactor, has several advantages, such as (i) it is not in contact with the sample (indirect sonication), thus reducing the contamination risk; (ii) allows to work with small amounts of samples and reagents; and finally (iii) presents low sample preparation times.

To the best of our knowledge, there is no in the literature that reports the use of high potency ultrasound-assisted cup-horn extraction in biological samples for metals determination related to human diseases, such as bipolar disorder. Thus, this study aimed to develop a fast and accurate method for the extraction of Zn, Cu, Fe, Li, Cd and Pb from human blood serum in order to differentiate patients with BD from other psychiatric disorders, such as schizophrenia.

3.3 Experimental

3.3.1 Materials and Reagents

3.3.1.1 Equipment

- Analytical balance (Shimadzu, model AX200, class I);
- Milli-Q purification system, model Quantum TM cartridge (Millipore, France);
- Ultracentrifuge, model Bio-Spin-R (BioAgency, Brazil);
- Inductively coupled plasma mass spectrometer (ICP-MS) (Shimadzu model 2030);
- Ultrasonic sonoreactor of cup-horn type (Qsonica®, Newtown, USA);
- Microwave oven, model DGT plus (Analytical).

3.3.1.2 Reagents

- HCl (37%, w/v, Merck);
- HNO3 (69%, w/v, Merck);
- Ultra-pure water with resistivity >18 MΩ obtained using Milli-Q system (Millipore, USA);
- Blood serum standard (Sigma Aldrich);
- Cu, Zn, Fe, Li, Cd and Pb (Scientific Hexis) standards.

3.3.1.3 Other Materials

- 15 and 50 mL polypropylene tubes;
- Glass materials for metal determination.

3.3.2 Sample Preparation

3.3.2.1 Acquisition of Samples

For this chapter, twenty-two serum samples were selected and classified into three groups: (i) control group; (ii) patients with SCZ; and (iii) patients with BD. The low number of samples, as well as the new classification of the study groups are justified in terms of the difficulty in obtaining new samples. As many collected samples were significantly used in the proteomic study, a few samples were left to be used in the ionomic study. In addition, another factor that influenced the low amount of sample was the samples characteristics, since we tried to select samples with similarities in

Table 3.1 Information of the volunteers involved in the research

	HC (n = 6)	BD (n = 8)	SCZ (n = 8)
Sex (male/female)	4/2	5/3	6/2
Age (year ± SD)	39 ± 16	36 ± 9	34 ± 9
Time of diagnostic (year ± SD)	–	5 ± 4	9 ± 8
Smoker (male/female)	0/1	2/2	2/1

SD Standard deviation

the profiles of the individuals involved. Table 3.1 summarizes the characteristics of the sample.

3.3.2.2   Ultrasound-Assisted Extraction Procedure

To evaluate the efficiency of ultrasonic assisted extraction, recovery experiments were performed using fortified blood serum with concentrations equal to $1840\,\mu g\,L^{-1}$ (Fe), $1600\,\mu g\,L^{-1}$ (Cu), $1270\,\mu g\,L^{-1}$ (Zn), $1.29\,\mu g\,L^{-1}$ (Pb), $5\,\mu g\,L^{-1}$ (Li) and $5\,\mu g\,L^{-1}$ (Cd). These concentration levels were based on certified materials charts, simulating the concentration of these elements in the blood serum. To ensure the interaction efficiency of the metals with the fortified sample with the metal, a waiting time of 15 min was expected. Afterwards, about 100 μL fortified blood serum was transferred to Lobind microtubes containing extracting solution composed of different concentrations of nitric acid (HNO_3) + hydrochloric acid (HCl). Then the mixture (sample + solution) was sonicated at room temperature using the high potency sonoreactor cup-horn type, varying the time (1, 3, 6, and 10 min) and the sonication amplitude (20, 40, 60, 80%).

After sonication, the supernatant liquid was separated from the solid phase by centrifugation for about 10 min at $13,000g$. Then, the supernatant was transferred to a cleaned Falcon tube and taken to ICP-MS analysis. For each series of extraction, a blank was also measured. The whites were prepared in the same way as the samples. The measurements were performed in triplicate for each extraction procedure. The external calibration method was used, and therefore the metal content calculations on samples were based on the calibration curve obtained from the standards. The percent recovery of all metals after ultrasonic extraction was calculated using the following equation:

$$\textbf{recovery } (\%) = (\textbf{obtained value/theoretical value}) \times \textbf{100}$$

3.3.2.3 Microwave-Assisted Decomposition Procedure

In microwave-assisted decomposition procedure using microtubes, 100 μL fortified blood serum were added to microtubes with a capacity of 2 mL. Then, 150 μL HNO_3 and 100 μL HCl were added. The sample decomposition program using the microwave oven was as follows: (i) 300 W for 20 min; and (ii) 0 W for 30 min. The microwave flasks containing the samples were then removed from the oven and kept in the hood for 30 min to wait for the cooling as well as the reduction of the pressure inside the flasks. After this waiting time, the vials were opened and the samples were transferred to Falcon-type polyethylene tubes with a capacity of 15 ml. The samples and their respective blanks were then added to 10 mL deionized water and analyzed by ICP-MS.

3.3.2.4 Determination Using ICP-MS

The elements determination in the serum sample after decomposition was performed with a ICP-MS (Shimadzu model 2030), equipped with a mini-torch, concentric neb-ulizer and refrigerated nebulization chamber with a constant temperature of 5 °C and assisted by a collision cell filled with helium (He). The Ion-lens voltage settings and other instrument parameters instrument were checked daily with solution containing beryllium (Be, 10 μg L^{-1}), indium, bismuth, cerium (In, Bi, Ce, 2 μg L^{-1}), cobalt and manganese (Co and Mn, 5 μg L^{-1}). Nebulizer, auxiliary and plasma (mini-torch) fluxes were 0.7, 1.10, and 8.0 L min^{-1}, respectively. The gas flow from the cell was 6.0 mL min^{-1}. The voltage of the lens was -21 V, and the energy filter was 7 V. Ana-lytical curves were prepared with concentration range from 0.1 to 3 μg L^{-1} for the elements Pb and Cd; and for the determination of Fe, Zn, and Cu, a analytical curve ranging from 10 to 100 μg L^{-1} was prepared. For Li, a calibration curve ranging from 0.5 to 10 μg L^{-1} was used.

3.4 Results and Discussion

3.4.1 Effect of the Extraction Solution Using Ultrasound Power

The composition of the extractive solution is an important parameter that signifi-cantly affects the extraction efficiency of metals in different samples. The sample nature and the acids properties used in the extraction are factors that influence the metals determination by ICP-MS [15]. For samples with high organic content, such as blood serum, HNO_3 is the recommended acid for the decomposition of such sam-ple, due to its oxidizing power [14]. Thus, to perform significantly the decomposition

Table 3.2 Recovery (%) of the elements after extraction by ultrasound using different concentration of HNO_3 (n = 3)

Element	Concentration of HNO_3 (%, ±RSD)		
	10% (v/v)	20% (v/v)	40% (v/v)
Zn	108 ± 1	118 ± 1	126 ± 2
Cu	100 ± 1	98 ± 2	108.0 ± 0.1
Fe	67 ± 4	72 ± 2	72 ± 1
Cd	88 ± 4	89 ± 3	90 ± 3
Pb	58 ± 1	54 ± 3	101 ± 3
Li	69 ± 3	76 ± 3	88 ± 3

of the serum, different concentrations of nitric acid were evaluated. The HNO_3 concentration ranged from 10 to 40% (v/v). For the initial experiment, sonic amplitude (60%) and sonication time (3 min) were kept constant. The final extraction volume of 350 μL was kept constant in all extraction studies using ultrasound.

Table 3.2 shows the results obtained after the extraction, using different concentrations of HNO_3. From Table 3.2, it was observed that the extraction efficiency for most metals increased with the increase of the acid concentration, except for the Fe element. The concentration of 40% (v/v) presented the best extraction (>80%) for the evaluated elements. In the 40% (v/v) concentration, the recovery of Pb (101 ± 3%, n = 3) is almost double when compared to the extractive solution containing 20% (v/v) HNO_3, where Pb recovery was 54 ± 3% (n = 3). However, interestingly, this solution containing 40% (v/v) of HNO_3 was not as efficient for Fe determination. The low recovery of Fe (<80%) may be justified due to the incomplete release of this element from the matrix. According to La Calle et al. [14], Fe is a problematic element to obtain good recoveries in most samples, when only nitric acid is used as the single component of the extractive solution, and suggests the use of HCl as an additive to potentiate the extraction of Fe, since there is the formation of Fe-chlorine complex ($Fe^{2+ ou 3+} + HCl \rightarrow FeCl_4^{2- ou 1-}$), facilitating its extraction. In this sense, this study also evaluated the synergy of HCl and HNO_3 as constituent of the extractive solution. For this, different concentrations of HCl (1–30%, v/v) were evaluated. The concentration of 40% (v/v) of HNO_3 was chosen as the optimal concentration for the decomposition of blood serum and used in the experiments to evaluate the influence of HCl as a component of the extractive solution.

The efficiency of the extractive solution using different concentrations of HCl mixed with nitric acid (40%, v/v) is shown in Table 3.3. From Table 3.3, it is observed that there was an increase in Fe recovery, as the HCl concentration also increases. The concentration of 30% (v/v) HCl present in the extractive solution allowed significant recovery of Fe (83.0 ± 0.2%, n = 3). It is important to highlight the maintenance of the recoveries of the other elements (>80%) using the extraction solution containing 30% (v/v) HCl + 40% (v/v) HNO_3.

Table 3.3 Recovery (%) of the elements after extraction by ultrasound using 40% (v/v) HNO_3 and different concentrations of HCl (n = 3)

Element	Concentration of HCl (%, ±RSD)			
	1% (v/v)	5% (v/v)	15% (v/v)	30% (v/v)
Zn	105 ± 1	104.0 ± 0.1	113 ± 4	115.0 ± 0.2
Cu	96.0 ± 0.4	96.0 ± 0.3	104 ± 4	98.0 ± 0.1
Fe	67 ± 2	70 ± 1	76 ± 1	83.0 ± 0.2
Cd	82 ± 1	83 ± 1	91 ± 3	96 ± 11
Pb	89 ± 2	89 ± 1	91 ± 3	105 ± 3
Li	88 ± 3	90 ± 2	91 ± 1	95 ± 2

3.4.2 Influence of the Sonication Time

The sonication time may influence the efficiency of analytes extraction in different samples, and therefore it was also evaluated. The sonication time ranged from 1 to 10 min. The analysis conditions were constant for: amplitude (60%), concentration of the extraction solution (HNO_3 (40%, v/v) + HCl (30%, v/v)), extraction solution volume (350 µL). The influence of the sonication time for recoveries of Zn, Cu, Fe, Cd and Pb metals after sonication are shown in Table 3.4. The results clearly show that the extraction efficiency of the elements increases with increasing sonication time. This fact is clearly visible when comparing the results obtained in the times of 1 and 3 min. A significant recovery is found for most elements when the sonication time is changed to 3 min. The recoveries ranged from 80 to 121%, with RSD varying between 0.2 and 6% (n = 3). While using the sonication time of 1 min, the recoveries ranged from 30 to 82% with RSD between 0.2 and 3% (n = 3). However, for the sonication times between 3 and 10 min, the recoveries do not change significantly and appear similar for most elements. In this way, 3 min was chosen as the time sufficient for the elements extraction, and therefore, it was used in the next experiments.

Table 3.4 Recovery (%) of the elements after extraction by ultrasound at different sonication times (n = 3)

Element	Sonication time (%, ± RSD)			
	1 min	3 min	6 min	10 min
Zn	70.0 ± 0.2	121.0 ± 0.3	115.0 ± 0.2	120 ± 1
Cu	53.0 ± 0.2	102.0 ± 0.2	98.0 ± 0.1	113 ± 1
Fe	30 ± 1	80 ± 4	83.0 ± 0.2	83 ± 1
Cd	58 ± 2	95 ± 6	96 ± 11	97 ± 2
Pb	82 ± 3	96 ± 6	105 ± 3	108 ± 3
Li	78 ± 4	90 ± 3	95 ± 2	98 ± 3

3.4.3   Influence of the Sonication Amplitude (Power)

The influence of the sonic power (amplitude) on the efficiency of analytes extraction from the blood serum was also evaluated. The studied sonication amplitude ranged from 20 to 80%, ranging in power from 70 to 210 W. The concentration of the extraction solution [HNO_3 (40%, v/v) + HCl (30%, v/v)], the volume of the extraction solution (350 μL), and the sonication time (3 min) were kept constant.

It is known that the ultrasound power transmitted to the medium is directly related to the vibration amplitude of the probe [15]. The influence of ultrasound amplitude on blood serum is shown in Table 3.5. Not surprisingly, the best recoveries for most metals are found with increasing amplitude, obtaining maximum recovery using the 80% amplitude (equivalent to 210 W power), where the recovery ranged from 83 to 121%, with RSD ranging from 0.1 to 4%. The significant improvement in recovery for most elements is evident when the amplitude is increased from 20% (78 W power) to 40% (107 W power). For a amplitude of 20%, the recovery ranged from 67 to 92%, with a RSD ranging from 1 to 4% (n = 3), while for the amplitude of 40%, a recovery between 76 and 99%, with RSD varying between 0.4 and 2% (n = 3). The results obtained for Zn, Cu, Li, Cd and Pb indicate that an amplitude of 40% is sufficient for a significant metals recovery, however due to the difficulty in recovering Fe with values higher than 80%, this study adopted the amplitude of 60% (160 W) as being the best amplitude to carry out the extractions of the elements of interest from the blood serum.

Table 3.5 Recovery (%) of the elements after extraction by ultrasound at different amplitudes of sonication (n = 3)

Element	Sonication amplitude (%, ±RSD)			
	20%	40%	60%	80%
Zn	88 ± 2	97 ± 1	121.0 ± 0.3	121.0 ± 0.2
Cu	92 ± 1	99 ± 1	102.0 ± 0.2	102.0 ± 0.1
Fe	67 ± 2	76 ± 1	80 ± 4	83.0 ± 0.1
Cd	83 ± 3	91.0 ± 0.4	95 ± 6	96 ± 2
Pb	89 ± 4	97 ± 1	96 ± 6	97 ± 1
Li	86 ± 3	90 ± 2	94 ± 2	96 ± 4

3.4.4 Comparison of the Efficiency Between the Proposed Method of Decomposition Using Ultrasound and the Conventional Method of Microwave-Assisted Digestion

To evaluate the efficiency of the proposed method using high power ultrasound, the technique of microwave assisted decomposition was used to check the accuracy. Thus, a comparison of the recovery results of Zn, Cu, Fe, Li, Cd, Pb using the serum decomposition using high power ultrasound (proposed method) and conventional microwave decomposition is presented in Table 3.6 where compares the recoveries of both decompositions with the concentrations of each metal used for the fortification of blood serum. From Table 3.6, there is significant agreement between the results obtained for ultrasonic assisted extraction and total assisted microwave digestion. Mean recoveries for metals were found in the range of 80–121% with RSD varying between 3 and 10% (n = 3) for the method using ultrasound, while for microwave oven decomposition, the recoveries ranged from 86 to 112%, with RSD ranging between 3 and 10% (n = 3). In relation to the total decomposition time of the sample, the extraction procedure using ultrasound showed to be an efficient, robust and fast method, since it presented results similar to those found with the microwave oven, however, using only about 15 min (3 min for sonication and 10 min for centrifugation) to prepare the sample, allowing a larger number of analyzes during one working day. While using microwave digestion, the decomposition procedure requires approximately 20 min (with a power of 300 W) to decompose the sample, 30 min to cool the reactor inside the microwave oven (with power 0 W) and approximately 30 min (at room temperature) to cool the reactor prior to opening, in addition to the time required to perform the reactor cleaning, thus restricting the sample preparation for a few analyzes during a working day. In view of these facts, the proposed method of ultrasound-assisted decomposition shows to be a robust, efficient and fast method for the determination of Zn, Cu, Fe, Li, Cd, Pb in blood serum and therefore will be used for determination of these elements in the blood serum of BD, SCZ, OD, HCF and HCNF in order to find differences in the concentration of such elements allowing to differentiate the BD from the other groups evaluated.

3.4.5 Proof of Concept: Evaluation of Metal Ion Concentration Differences in Serum Samples from Patients with Bipolar Disorder, Schizophrenia and Healthy Individuals

Several studies have associated some psychiatric disorders with the metal deregulation from body fluids and brain tissues [20, 22, 24]. However, little is known about the difference in concentration of Zn, Cu, Li, Fe, Cd and Pb in serum samples from

Table 3.6 Comparison of the efficiency of the proposed method using ultrasound with conventional decomposition using microwave oven ($p < 0.05$)

Element	Reference value (μg mL^{-1})	Measured value (n = 3; μg mL^{-1})		Recovery (n = 3; %)		Calibration curve (μg L^{-1})	R^2	LD (μg L^{-1})	LQ (μg L^{-1})
		Ultrasound	Microwave	Ultrasound	Microwave				
Fe	1.84	1.468 ± 0.059	1.578 ± 0.047	80 ± 4	86 ± 3	10–100	0.992	0.1	0.3
Cu	1.6	1.62 ± 0.11	1.442 ± 0.072	102 ± 7	90 ± 5	10–100	0.999	0.02	0.05
Zn	1.27	1.53 ± 0.15	1.422 ± 0.057	121 ± 10	112 ± 4	10–100	0.999	0.1	0.2
Pb	0.00129	0.00126 ± 0.00004	0.00126 ± 0.0001	97 ± 3	98 ± 8	0.1–3	0.999	0.01	0.03
Cd	0.005	0.0047 ± 0.0003	0.0054 ± 0.0005	95 ± 6	109 ± 10	0.1–3	0.999	0.002	0.01
Li	0.005	0.0045 ± 0.0003	0.0048 ± 0.0004	90 ± 7	96 ± 8	0.5–10	0.999	0.01	0.04

Fig. 3.2 Comparison of metal concentration ($\mu g\,L^{-1}$) in bipolar disorder (BD, black), healthy control (CS, blue) and schizophrenia (SCZ, red). Statistical analyzes were performed using t-test (n = 3), where $*p < 0.01$; $***p > 0.05$

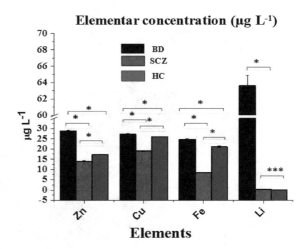

patients with BD, patients with SCZ, and healthy individuals. Thus, to understand the influence of such metal ions on the pathophysiology of BD and SCZ, the proposed method of ultrasound-assisted extraction was applied to the blood of the group evaluated to extract analytes of interest and to try to understand the possible metallomic profile difference between BD and healthy individuals as well as schizophrenia.

After the application of the extraction method proposed as sample preparation procedure, and the application of ICP-MS as analysis technique in the serum samples of each group, the absence of Pb and Cd was observed for all the samples evaluated. However, the Zn, Cu, Li, and Fe elements were significantly associated with such mental disorders (Fig. 3.2).

The non-determination of Pb and Cd in the samples evaluated can be justified in terms of environmental contamination. In other words, the exposure of the individuals evaluated to such metals is so insignificant that the present levels of Pb and Cd are below the limit of quantification presented by the methodology, thus justifying the non-determination of such elements. The absence of Pb and Cd in all samples evaluated suggests that both metals may not be involved in the pathophysiology of SCZ and BD.

For the other metals, there is a contrast in the levels of all metals found for BD and SCZ when compared to healthy individuals. Interestingly, for the group of patients with BD, all metal ions were found at higher levels than the SCZ and HC groups (Fig. 3.2), while for the group of patients with SCZ, it was observed that all the metal ions evaluated, except Li, were found at lower levels (Fig. 3.2). The PCA plots of the first components show a significant separation between the evaluated groups, especially between BD patients and HC individuals, as well as SCZ patients. However, when the second component is observed, a small separation between SCZ and HC patients is found (Fig. 3.3). From Fig. 3.4, it is observed that Cu is the main component for the significant separation of the BD group from the other groups evaluated. For the SCZ groups, Zn is responsible for such separation, while for the HC group, Fe performs this function.

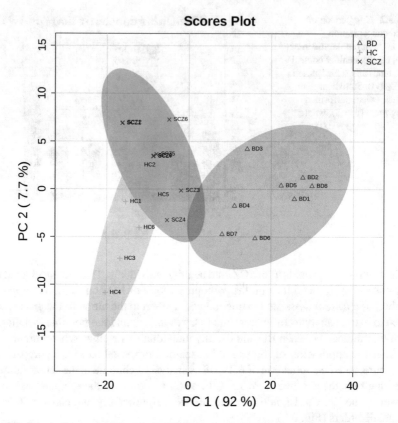

Fig. 3.3 Principal components analysis (PCA) for discrimination of patients with bipolar disorder (BD), patients with schizophrenia (SCZ) and healthy control (HC). Each symbol represents an individual patient and the corresponding spatial distribution of these symbols reveals similarities and differences between them

Zinc and copper play an important role in a variety of biochemical processes that modulate the function of the immune and central nervous systems. For example, the change in blood zinc homeostasis may reflect mood disorders. In agreement with this statement, our results showed that, for the group of patients with BD, the levels of Zn increased, while for the groups of SCZ, the concentration of Zn was lower. Similar results are described in the literature. For example, Cai et al. [8] reported a decrease in serum Zn levels of patients with SCZ, suggesting that Zn may be involved in SCZ pathophysiology. Sussulini et al. [28] also reported an increase of the Zn level in serum of patients with BD when compared to healthy control.

Our results also show a decrease in serum Cu levels in patients with SCZ, and an increase in serum levels of patients with BD. Cu deregulation in the biological system may contribute to the pathophysiology of brain diseases. This fact is justified because Cu is an essential element for the activity of a series of important enzymes such as cytochrome C oxidase, dopamine-beta-hydroxylase and Cu/Zn superoxidase

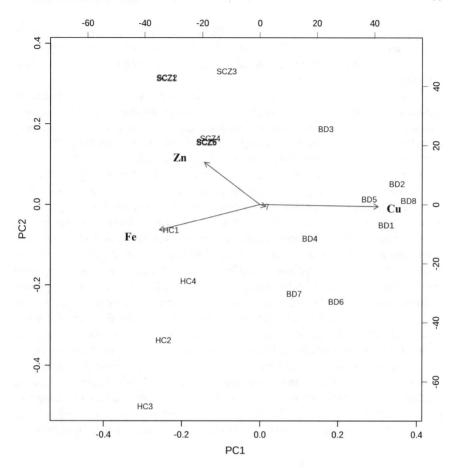

Fig. 3.4 Profile of the elements for schizophrenia (SCZ), bipolar disorder (BD) and healthy control (HC). The figure shows the main components for the significant separation between the groups

dismutase, which are critical in the elimination of reactive oxygen species [27]. Failures related to any of these enzymes due to decrease or increase in Cu may explain the origin of the mental diseases. In one study, Liu et al. [17] found lower concentration of Cu in the serum of patients with SCZ and associated the decrease in the concentration of the element with risk of schizophrenia. While Trasobares et al. [31] found higher levels of Cu in the serum of patients with BD, associating this increase with the BD pathophysiology.

In addition, our study also demonstrated a significant increase of iron in serum of patients with BD, as well as a decrease of the Fe concentration in serum of schizophrenics when compared to the group of healthy individuals. These results corroborate those reported by Mustak et al. [20] and Liu et al. [17] which showed differences in Fe concentration in serum samples of BD and SCZ.

Regarding Li, without surprises, a significant increase was observed in the levels of this metal in blood serum of patients with BD. This fact is justified because Li is the main agent in the treatment of mood disorder, especially bipolar disorder. Thus, our results can be considered satisfactory.

A very strong relationship can be established between the results obtained in this ionomic study with those results reported in the proteomic study (Chap. 1). For example, in this study significant differences were observed in iron levels for both diseases (BD and SCZ). Iron is essential for normal neurological function, effectively participating in the synthesis of myelin and neurotransmitters [23]. Studies have reported the accumulation of iron ion in patients' brains with neurological diseases, and iron is one of the responsible for the production of free radicals and oxidative stresses in the biological activities of the brain [23, 25]. For the studies, increased iron ion concentration in the brain of patients with neurological diseases is also directly related to increased transferrin protein in some regions of the brain. This information corroborates the results found in this thesis, since the elevation of transferrin (see Table 1.3, Chap. 1) and free iron ions (Fig. 3.2, this chapter) was observed in blood serum of patients with BD when compared healthy individuals. Another important relation between the two studies (proteomic and ionomic) is related to the zinc and copper ions found significantly altered in this thesis study for BD and SCZ when compared to the control. Both metal ions (Zn and Cu) can interact with amyloid proteins, such as serum amyloid P component, forming aggregates. Such aggregates may be related to severe symptoms triggered by neuropsychiatric disorders, such as BD and SCZ [1, 7]. An increase in zinc level (Fig. 3.2, this chapter) and serum amyloid P protein components (Table 1.3, Chap. 1) has been observed in a sample of patients with BD when compared to patients with schizophrenia, showing thus direct relationship between proteins and metals in the pathophysiology of bipolar disorder.

3.5   Partial Conclusion

The proposed method based on ultrasonic assisted extraction provides a rapid and efficient sample preparation for the determination of Zn, Cu, Fe, Li, Cd and Pb in human serum samples by ICP-MS discriminating patients with bipolar disorder from other psychiatric diseases. The cup-horn sonoreactor proved to be efficient for the extraction of metal together with the extraction solution containing 40% (v/v) HNO_3 + 30% (v/v) HCl with short sonication times (3 min) and low sonication amplitude (60%). The ability to perform simultaneous extractions from several samples makes this ultrasonic system advantageous compared to microwave assisted digestion. Furthermore, after the application of the proposed method in blood serum samples from patients with BD, SCZ and healthy individuals, significant differences were observed in Fe, Cu, Zn and Li levels for BD and SCZ when compared to the controls. For the

BD group, metals were observed at a high level, while for the SCZ group, all metals were found at low levels. These results may help to understand the biochemical behavior of metals in the pathophysiology of psychiatric diseases, specifically bipolar disorder.

References

1. Adlard PA, Bush AI (2018) Metals and Alzheimer's disease: How far have we come in the clinic? J Alzheimer's Dis 62:1369–1379
2. Agatemor C, Beauchemin D (2011) Matrix effects in inductively coupled plasma mass spectrometry: a review. Anal Chim Acta 706(1):66–83
3. Ahmed SS, Santosh W (2010) Metallomic profiling and linkage map analysis of early parkinson's disease: a new insight to aluminum marker for the possible diagnosis. PLoS ONE 5:1–6
4. Batista BL, Rodrigues JL, Nunes JA, Souza JA, Souza VC, Barbosa F Jr (2009) Exploiting dynamic reaction cell inductively coupled plasma mass spectrometry (DRC-ICP-MS) for sequential determination of trace elements in blood using a dilute-and-shoot procedure. Anal Chimica Acta 639:13–18
5. Bocca B, Alimonti A, Petrucci F, Violante N, Sancesario G, Forte G, Senofonte O (2004) Quantification of trace elements by sector field inductively coupled plasma mass spectrometry in urine, serum, blood and cerebrospinal fluid of patients with parkinson' s disease. Spectrochimica Acta Part B Atom Spectrosc 59:559–566
6. Bocca B, Forte G, Petrucci F, Senofonte O, Violante N, Alimonti A (2005) Development of methods for the quantification of essential and toxic elements in human biomonitoring. Annali dell'Istituto Superiore Di Sanita 41(2):165–170
7. Bush AI (2013) The metal theory of alzheimer's disease. Rev Lit Arts Am 33:277–281
8. Cai L, Chen T, Yang J, Zhou K, Yan X, Chen W, Sun L, Li L, Qin S, Wang P, Yang P, Cui D, Burmeister M, He L, Jia W, Wan C (2015) Serum trace element differences between schizophrenia patients and controls in the han chinese population. Sci Rep 10:1–8
9. Gajek R, Barley F, She J (2013) Determination of essential and toxic metals in blood by ICP-MS with calibration in sybthetic matrix. Anal Methods 2193–2202
10. Garcia JS, Magalhães CS, Arruda MAZ (2006) Trends in metal-binding and metalloprotein analysis. Talanta 69:1–15
11. Heitland P, Köster HD (2006) Biomonitoring of 37 trace elements in blood samples from inhabitants of northern germany by ICP-MS. J Trace Elem Med Biol 20(4):253–262
12. Ivanenko NB, Ivanenko AA, Solovyev ND, Zeimal AE, Navolotski DV, Drobyshev EJ (2013) Biomonitoring of 20 trace elements in blood and urine of occupationally exposed workers by sector field inductively coupled plasma mass spectrometry. Talanta 116:764–769
13. Jakubowski N, Moensb L, Vanhaeckeb F (1998) Sector field mass spectrometers in ICP-MS. Spectrochimica Acta Part B 53:1739–1763
14. La Calle I, Costas M, Cabaleiro N, Lavilla I, Bendicho C (2012) Use of high-intensity sonication for pre-treatment of biological tissues prior to multielemental analysis by total reflection X-ray fluorescence spectrometry. Spectrochimica Acta Part B 67:43–49
15. Lavilla I, Costas M, Gil S, Corderi S, Sanchez G, Bendicho C (2012) Simplified and miniaturized procedure based on ultrasound-assisted cytosol preparation for the determination of Cd and Cu bound to metallothioneins in mussel tissue by ICP-MS. Talanta 93:111–116
16. Lin T, Liu T, Lin Y, Yan L, Chen Z, Wang J (2017) Comparative study on serum levels of macro and trace elements in schizophrenia based on supervised learning methods. J Trace Elem Med Biol 43:202–208
17. Liu T, Lu QB, Yan L, Guo J, Feng F, Qiu J, Wang J (2015) Comparative study on serum levels of 10 trace elements in schizophrenia. Plos One 1–8

18. Marchante-Gayón JM, Muniz CS, Alonso JIG, Sanz-Medel A (1999) Multielemental trace analysis of biological materials using double focusing inductively coupled plasma mass spectrometry detection. Anal Chim Acta 400:307–320
19. Mounicou S, Szpunar J, Lobinski R (2009) Metallomics: the concept and methodology. Chem Soc Rev 3:1119–1138
20. Mustak MS, Rao TS, Shanmugaveleu P, Sundar NM, Menon RB, Rao RV, Rao KS (2008) Assessment of serum macro and trace element homeostasis and the complexity of inter-element relations in bipolar mood disorders. Clin Chimica Acta 394:47–53
21. Palmer CD, Lewis ME Jr, Geraghty CM, Barbosa F Jr, Parsons PJ (2006) Determination of lead, cadmium and mercury in blood for assessment of environmental exposure: a comparison between inductively coupled plasma—mass spectrometry and atomic absorption spectrometry. Spectrochimica Acta Part B 61:980–990
22. Pfaender S, Grabrucker AM (2014) Characterization of biometal profiles in neurological disorders. Metallomics 6:960–977
23. Pinero DJ, Connor JR (2000) Iron in the brain: an important contributor in normal and diseased states. Neuroscientist 6:435–453
24. Roos PM, Lierhagen S, Flaten TP, Syversen T, Vesterberg O, Nordberg M (2012) Manganese in cerebrospinal fluid and blood plasma of patients with amyotrophic lateral sclerosis. Experim Biol Med 237:803–810
25. Sayre LM, Perry G, Harris PL, Liu Y, Schubert KA, Smith MA (2000) In situ oxidative catalysis by neurofibrillary tangles and senile plaques in Alzheimer's disease. J Neurochem 74(1):270–279
26. Shi W, Chance MR (2011) Metalloproteomics: forward and reverse approaches in metalloprotein structural and functional characterization. Curr Opin Chem Biol 15(1):144–148
27. Siwek M, Styczen K, Sowa-Kucma M, Dudek D, Reczynski W, Szewczyk B, Misztak P, Opoka W, Topor-Madry R, Nowak G, Rybakowski JK (2016) The serum concentration of copper in bipolar disorder. Psychiatr Pol 2674:1–13
28. Sussulini A, Banzato CEM, Arruda MAZ (2011) Exploratory analysis of the serum ionomic profile for bipolar disorder and lithium treatment. Int J Mass Spectrom 307(1–3):182–184
29. Szpunar J (2004) Metallomics: a new frontier in analytical chemistry. Anal Bioanal Chem 378(1):54–56
30. Thomas R (2008) Practical guide to ICP-MS: a tutorial for beginners, 2nd edn. Taylor & Francis Group, Boca Raton, pp 1–339
31. Trasobares EM, Tajima K, Cano S, Fernández C, Luis J, Unzeta B, Arroyo M, Fuentenebro F (2011) Trace elements in bipolar disorder. J Trace Elements Med Biol 255:78–83
32. Vogiatzis CG, Zachariadis GA (2014) Tandem mass spectrometry in metallomics and the involving role of ICP-MS detection: a review. Anal Chim Acta 819:1–14
33. Wolle MM, Rahman GM, Kingston HM, Pamuku M (2014) Speciation analysis of arsenic in prenatal and children's dietary supplements using microwave-enhanced extraction and ion chromatography—inductively coupled plasma mass spectrometry. Anal Chim Acta 818:23–31

Chapter 4
Development of a New Polymer Membrane Based Methodology for the Characterization of the Urine Proteome

4.1 Objectives

4.1.1 General Objective

Carry out a study of the protein recovery extracted from urine samples, using a simple and robust methodology of filtration based on polymer membranes. This stage of the thesis was developed during my mission in the Faculty of Sciences and Technology of the New University of Lisbon, Portugal, under the supervision of Prof. Dr. José Luis Capelo Martinez.

4.1.2 Specific Objectives

- Develop a new simple and robust methodology to recover proteins from the urine, making use of polymer membranes;
- Evaluate the optimal conditions to extract proteins from urine, which are:
  (i)   composition of the membranes (cellulose, nitrocellulose and cellulose acetate);
  (ii) pores of the membranes (0.22 and 0.45 μm);
  (iii) flow rate (0.25, 0.5 and 1.0 mL min^{-1});
  (iv) pH of the medium (3–7).

- Carry out protein extractions, using the best evaluated conditions;
- Apply the methodology developed in a gender classification study.

© Springer Nature Switzerland AG 2019
J. R. de Jesus, *Proteomic and Ionomic Study for Identification of Biomarkers in Biological Fluid Samples of Patients with Psychiatric Disorders and Healthy Individuals*, Springer Theses, https://doi.org/10.1007/978-3-030-29473-1_4

4.2   Review of the Literature

4.2.1   Urine as a Biomarkers Source of Human Diseases

Human urine together with blood are one of the major biological samples that play a key role in clinical diagnostics [12]. Because the urine is a non-invasive sample, its proteome have been extensively investigated in recent years to find differential proteins with a potential biological marker of different human diseases [1, 9].

Urine is produced by the kidney, which is made up of thousands of functional cells called nephrons. The nephrons are divided into glomerulus and renal tubules, the former filtering the blood originating the primitive urine and the latter absorbing this primitive urine [11]. The final urine arrives through the ureters to the bladder, where it is stored until elimination (Fig. 4.1). As the final destination of many biomolecules involved in biochemical reactions, urine allows obtaining information from different organs and processes, and can be an important source for studies related to human diseases [2].

The biomarkers identification through urine can lead to the development of simple diagnostic tests for important human diseases, as well as allowing the choice of the best therapeutic treatment [14]. However, although the concentration of some proteins is adequate for mass spectrometry analysis, the sample treatment for urine-based proteomic analysis has a number of challenges [9]. Examples such as the dilution of protein concentration and the high content of salts present in urine are factors that affect urinary proteome analysis, interfering the detection of candidate proteins for biological markers. In addition, because proteomic analyzes are generally performed after long period of urine collection, the instability of many proteins, especially those of lower abundance, may interfere with the results obtained from a longitudinal proteomic study [7]. Therefore, sample preparation procedures are necessary in order to minimize/eliminate interferers in the proteomic analysis of the urine or even to isolate proteins of interest [8].

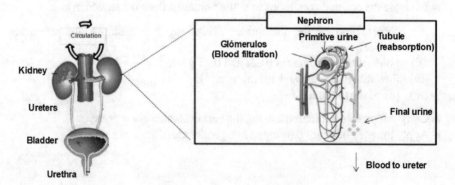

Fig. 4.1 Illustration of the urine formation by the kidneys

4.2.2 Sample Preparation for Characterization of the Urine Proteome

In the literature, several protocols employed to isolate and/or concentrate urine proteins are described. For example, precipitation of protein using organic solvents is the simplest and most widely used methodology for studying proteomics involving urine [3, 8]. In this methodology, various solvents are applied as a mixture in different proportions, or pure. Among the most commonly used solvents are: acetone, methanol, chloroform, dichloromethane and trichloroacetic acid (TCA). After precipitation using the solvents, the samples are centrifuged and the supernatant and pellet are separated. Depending of the solvent chemical properties used and the proportions applied, proteins with specific characteristics, such as hydrophobicity or hydrophilicity may be precipitated. Thongboonkerd et al. [13], for example, reported the application of acetone for precipitation of hydrophilic proteins. To precipitate hydrophobic proteins, the researchers used chloroform and methanol.

Another methodology applied in a study of urinary proteome characterization is dialysis followed by lyophilization. This methodology combines two steps in sequence. In the first step, the urine samples suffer dialysis, generally using a 3.5 kDa cutoff membrane, after that the solution resulting from the dialysis is transferred to a lyophilizer where the evaporation is carried out. This method represents one of the most efficient methodologies in the significant extraction of urine proteins [13].

Ultracentrifugation also has been reported as an important methodology applied to characterize the proteome of urine. The methodology is based on the application of the centrifugal force in a sedimentation experiment, where the denser components are separated from the other components. For protein extraction, a centrifugal filter with molar mass cutoff of interest is used. Sigdel et al. [10], for example, used a 5 kDa cutoff filter to monitor renal transplanted patients.

Although all these methodologies present relative advantages for the extraction of proteins from the urine, most of them present numerous limitations. For example, in relation to the ultracentrifugation, the main limitation is observed in the selectivity of the extracted proteins, once this methodology presents good efficiency only in the extraction of proteins with basic and hydrophobic characteristics [14]. Regarding the methodology using dialysis followed by lyophilization, the main limitations are the long time required in the dialysis and lyophilization procedure, as well as the high concentration of salts resulting from the evaporation of the liquid, which can significantly interfere in the analysis by mass spectrometry [13]. Given these limitations, information on the effectiveness of these different methods in terms of quality and recovery performance is questionable. In addition, one of the main questions about these methodologies is how reliable is the result obtained from the proteomic study with proteins stored at long periods, where many proteins can degrade.

Thus, the objective of this thesis chapter was to develop a simple and robust methodology based on a polymer membrane that allows the total extraction of the proteins from the urine, in order to establish a quantitative and qualitative study of proteins extracted from the urine for the detection of proteins that are candidates for biomarkers of human diseases.

To perform the extraction of urine proteins, cellulose-based polymer membranes with different compositions (cellulose, CE, nitrocellulose, NC and cellulose acetate, CA) and pore sizes (0.22 and 0.45 μm) were tested. In addition, parameters such as pH of the medium (3, 4, 5, 6 and 7), urine flow (0.25, 0.5 and 1.0 mL min^{-1}) passing through of the membrane was also evaluated.

4.3 Experimental

4.3.1 Materials and Reagents

4.3.1.1 Equipment

- Analytical balance (Shimadzu, model AX200, class I);
- Direct current source (Armersham Biosciences, Sweden);
- Matrix-assisted laser ionization/desorption mass spectrometer (MALDI-TOF MS) model Ultraflex II (Bruker Daltonics);
- Liquid chromatography coupled to the mass spectrometer (nLC-MS/MS) EASY-nLC II (Bruker Daltonics);
- Scanner, Image Scanner™ II model (GE Healthcare, Sweden);
- SDS-PAGE electrophoresis system, model SE 600 Ruby (GE Healthcare, Sweden);
- Ultracentrifuge, model Bio-Spin-R (BioAgency, Brazil).

4.3.1.2 Buffer and Solutions

- Dye solution (colloidal coomassie): 10% (v/v) phosphoric acid, 10% ammonium sulfate (w/v), comassie 0.12% G-250, 20% (v/v) methanol, deionized water;
- Gel solution: 30% acrylamide (w/v), N,N-methylenebisacrylamide 0.8% (w/v), deionized water;
- Extractive solution—Triton X-100 1% (v/v), and SDS 2% (w/v), Tris-HCl (50 mmol L^{-1}, pH 9.0);
- Tris-HCl buffer pH 8.8: deionized water, Tris-base 1.5 mol L^{-1}, pH adjustment with hydrochloric acid (HCl);
- Rehydration buffer: urea (7 mol L^{-1}), ammonia bicarbonate (12 mmol L^{-1}) deionized water;
- Sample Buffer: Tris-HCl pH 6.8 (40 mmol L^{-1}), 10% (v/v) β-mercaptoethanol, 50% (v/v) glycerol, 10% SDS (w/v), bromophenol 0.1% (w/v);
- (10-fold concentrate): Tris-base (2 mmol L^{-1}), glycine (192 mmol L^{-1}), 0.1% SDS (w/v), deionized water.

4.3.2 Methods

4.3.2.1 Acquisition of Samples

For this study, urine collection was performed according to the international urine collection protocol developed by the European Kidney and Urine Proteome (EuroKUP) and Human Kidney and Urine Proteome Project (HKUPP) [16]. Briefly, the second morning urine of eight healthy subjects (5 males com age of 26 ± 3 years and three females with age 25 ± 3 years) were collected with sodium azide 1 mmol L^{-1} and maintained at 4 °C. The collected urine was pretreated by centrifugation $2500 \times g$ for 20 min to remove cells, bacteria and other debris. Afterwards, the urines were grouped respecting the male and female groups. Aliquots of 20 mL were stored at -60 °C until further analysis.

4.3.2.2 Solid Phase Extraction Procedure of Urine Proteins Using Cellulose-Based Membrane

For the study of urine protein extraction, a series of optimizations were performed. All experiments were performed in triplicate. A general scheme of the whole experimental procedure is presented in Fig. 4.2.

Following, all evaluated parameters are displayed:

(i)   Type of membranes (cellulose acetate, nitrocellulose and cellulose);
(ii) Pores of the membranes (0.22 and 0.45 μm);
(iii) Urine flow (0.25, 0.5 and 1.0 mL min^{-1});
(iv) pH of the medium (3–7).

To ensure a greater interaction between the proteins and the membrane, some parameters were pre-defined, for example, two cut membranes were installed inside syringe, and the procedure of passing the urine through the membrane was repeated three times. The membranes (cellulose, nitrocellulose and cellulose acetate) were disc cut to a diameter of approximately 1.2 cm, and then carefully placed into a 5 mL medical syringe. The complete syringe outlet seal was required to maximize the yield of the adsorption proteins/membranes. For this, optionally, medical gauzes were inserted into the syringe outlet prior to the installation of the membranes. In order to have a greater reproducibility of the replicates and to control the urine flow (0.25, 0.5 and 1.0 mL min^{-1}) through the membrane, a solid phase extraction system (SPE) was used. The salt and the possible contaminants were eliminated of the membrane by the addition of 1 mL deionized water at the end of the process.

For one-dimensional electrophorese analysis (SDS-PAGE), the urine proteins were eluted from the membrane using an extraction solution composed of Triton X-100 1% (v/v) and SDS 2% (w/v), Tris-HCl (50 mmol L^{-1} pH 9.0). In general, the membrane with the adsorbed urine proteins were cut into small pieces and inserted into a new clean microtube. Then, 0.3 mL of the extractive solution was added. The

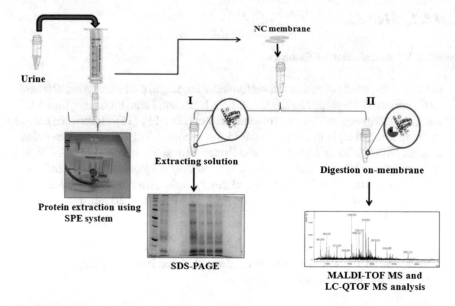

Fig. 4.2 Urine sample preparation. Using a solid phase extraction system (SPE), the urine is passed through a cartridge containing a piece of gauze to completely seal the syringe, and two polymer membranes (1.2 cm diameter) that will serve to retain the proteins from the urine. In stage I, the proteins are extracted from the polymer membrane and the one-dimensional electrophoreses (SDS-PAGE) is performed to evaluate the efficiency of protein extraction. In step II, the proteins adsorbed on the polymer membranes are digested using trypsin and then analysis is performed by MALDI-MS/MS and nLC-MS/MS

tube containing the extractive solution together with the membrane was homogenized for 10 min at room temperature and then stirred for 15 min in an ultrasonic bath. The supernatant was transferred to a new microtube, and subjected to evaporation to dryness in a vacuum concentrator centrifugal without heating. The remaining urine proteins, after evaporation, were reconstituted in 25 μL of a solution containing urea (8 mol L^{-1}), CHAPS (2%, w/v), thiourea (3 mol L^{-1}). After the extraction process, the concentration of proteins present in the membrane was measured using Bradford assay.

Each experiment was performed in triplicate for quality control and statistical analysis. The following equation describes the calculation performed to evaluate the extraction efficiency (E):

$$\mathbf{E} = (\mathbf{Ci} - \mathbf{Cf}) \times \mathbf{100\%}$$

where Ci is the concentration of protein (μg μL^{-1}), present in the urine before passing through the membrane, and Cf is the concentration of protein (μg μL^{-1}) after passing through the membrane.

4.3.2.3 Proteins Separation by SDS-PAGE

Approximately 30 μg of proteins resulting from the extraction of the membranes were mixed with sample buffer (see Sect. 4.3.1.2). Proteins were denatured by heating (98 °C for 3 min) and then proteins were loaded onto 12% (v/v) polyacrylamide gel. The electrophoretic separation was performed under the following conditions: voltage of 200 V; current 40 mA and power 7 W for 1 h. After electrophoretic separation, the gel was stained with coomassie blue to reveal the protein bands. Images of the gels were obtained using an Image Scanner.

4.3.2.4 On-Membrane Protein Digestion

In order to obtain a total digestion of the urine proteins adsorbed on the nitrocellulose membrane, trypsin was applied directly to the membrane, resulting in a differential procedure of this study. For this, the membrane with the adsorbed proteins were cut into small pieces and then inserted into a new microtube. The disulfide bonds of urine proteins were reduced using 0.3 mL dithiothreitol (DTT, 10 mmol L^{-1}, solubilised in ammonium bicarbonate (Ambic), 12.5 mmol L^{-1}). The samples were homogenized and incubated for 60 min at 37 °C. The resulting cysteines were alkylated using 0.3 mL of iodoacetamide (IAA) (50 mmol L^{-1}, solubilized in Ambic, 12.5 mmol L^{-1}) for 45 min at room temperature. Then, 0.3 mL of Ambic were added to remove any trace of IAA that could interfere with the action of trypsin during the cleavage process. Immediately after sample washing, using Ambic (12.5 mmol L^{-1}), trypsin (0.4 μg μL^{-1}) was added, respecting the ratio of 1:6 (trypsin: protein, w/w). Once the trypsin was added, the on-membrane digestion was performed overnight at 37 °C. After that, the resulting peptides were transferred to a new tube, and 6 μl of formic acid (99%, v/v) were added in order to finish the enzymatic activity. Finally, the digested proteins were dried and reconstituted in 0.1 mL (0.1%, v/v), and then desalted using Pierce® C18 Tips.

4.3.2.5 Analysis by MALDI-TOF MS

After on-membrane tryptic digestion, the peptides obtained were dried and reconstituted into formic acid (0.3%, v/v). Then 0.5 μL of the sample and 1 μL of the α-cyano-4-hydroxycinnamic acid solution (7 mg mL^{-1}, solubilised in trifluoroacetic acid, TFA, 0.1% v/v, and ACN, 50% v/v) were applied to MALDI plate and then analyzed by MALDI-TOF MS Ultraflex II model (Bruker Daltonics). The mass spectrometer was operated in positive ion mode. The spectra were acquired in the range of m/z between 600 and 3500. A total of 500 spectra were acquired for each sample at a frequency of 50 Hz. External calibration was performed with the monoisotopic $[M + H]^+$ peaks of bradykinin 1 (m/z 757.3992), angiotensin II (m/z 1046.5418), angiotensin I (m/z 1296.6848), substance P (m/z 1758.9326), ACTH clip 1-17 (m/z 2093.0862), ACTH 18-39 (m/z 2465,1983) and somatostatin 28 (m/z 3147.4710).

The search for identification of peptides was searched with the MASCOT search programs with the following parameters: (i) SwissProt Database; (ii) protein molecular mass: all; (iii) a cleavage loss; (iv) fixed modifications:carbamidomethylation (C); (v) variable modifications: oxidation of methionine and (vi) tolerance up to 50 ppm after external calibration.

4.3.2.6 Analysis by NLC-MS/MS

Analysis by nLC-MS/MS was performed using an EASY-nLC II instrument (Bruker Daltonics), equipped with a pre-column EASYColumn, C18-A1, 2 cm, internal diameter 100 μm (Thermo Fisher Scientific) and a column C18-A2, 10 cm, 75 μm internal diameter (Thermo Fisher Scientific). Chromatographic separation was performed on a buffer B gradient (90% v/v ACN and 0.1% v/v formic acid) at a flow rate of 300 nL min^{-1} for 90 min. All spectra were acquired in the range of 150–2200 Da. LC-MS/MS data were analyzed using Software Data Analysis 4.2 (Bruker). Proteins were identified using Mascot (Matrix Science, UK). The MS/MS spectra were searched against the Swissprot 57.15 database (515,203 sequences, 181,344,896 residues), establishing the taxonomy for Homo sapiens (20,266 sequences). The data of Tandem MS were investigated using MASCOT, with the following parameters: precursor mass tolerance of 20 ppm, tolerance to the 0.05 Da fragment, trypsin specificity with a maximum of 2 lost cleavages, cysteine carbamidomethylation defined as modification fixed and oxidation of methionine and acetyl (N-term) as a variable modification. The significance threshold for the identifications was set at $p < 0.05$.

4.3.2.7 Statistical Analysis

For statistical analysis, five replicates of each sample were applied to the MALDI plate. Data from the corresponding gross spectra of each sample were preprocessed using the Mass-Up program (http://sing.ei.uvigo.es/mass-up/). Then, the principal component analysis (PCA) was performed with the following parameters: (i) maximum components (-1 for no maximum number of components); (ii) coverage variation (0.95); (iii) normalize. A hierarchical clustering analysis was also applied as a complement to the PCA analysis. For group analysis the following parameters were used: (i) minimum variance (0); (ii) minimal intra-sample presence (0); (iii) deep grouping (no).

4.4 Results and Discussion

4.4.1 Optimization of Parameters for Extraction of Urine Proteins Using Polymer Membranes

4.4.1.1 Influence of Membrane Composition

To discuss the effect of membrane morphology on the efficiency of urine protein extraction, three types of cellulose-based membranes (cellulose, CE, nitrocellulose, NC, and cellulose acetate, CA) were evaluated. For this, membranes with the same pore size ($0.22 \, \mu m$) were applied. In addition, the experiment was performed using an initial urine flow rate of $1 \, mL \, min^{-1}$ and a pH value of 7.0. In order to obtain increased protein-membrane interaction, two cut membranes were installed in the syringe, and urine was passed through the membrane three times. Figure 4.3, panel A, illustrates the results obtained. From Fig. 4.3, it was observed that, among the membranes used, NC was the membrane that presented the best results, with protein recovery of $68 \pm 3\%$ ($n = 3$), while CA and CE allowed to extract about $61 \pm 3\%$ ($n = 3$) and $59 \pm 1\%$ ($n = 3$), respectively. This fact can be justified in terms of the chemical properties

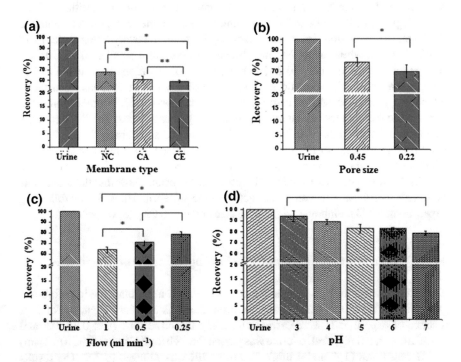

Fig. 4.3 Evaluation of the best conditions for the extraction of urine proteins, using polymer membrane. **a** membrane composition; **b** pore size; **c** urine flow; **d** influence of pH. Statistical analysis was performed using t-test, where $*p < 0.05$ and $**p > 0.05$

of the polymers [6]. Different membrane-forming materials have different intrinsic affinity attraction capabilities for different proteins. Thus, the adsorption of proteins on the membrane depends on various chemical interactions, including electrostatic attraction, van der Waal interaction, hydrophobic interaction and hydrogen bonding. Although the three membranes are capable of carrying hydrogen bonding due to the oxygen molecules present in their structure, the strong dipole of the nitrate group on the NC membrane interacts with the strong dipole of the peptide bond in the protein, giving rise to a strong affinity electrostatic. In this case, it is expected that the electrostatic interaction between the membrane and the protein is the main factor that dominates the adsorption behavior, justifying, therefore increase of the protein recovery by the NC membrane in relation to the CA and CE [4]. Thus, considering the results presented above, the NC membrane was selected to perform the following experiments.

4.4.1.2   Influence of Pore Size

To evaluate the effect of membrane pore size on urine protein extraction, two different pore sizes of the NC membrane (0.22 and 0.45 μm) were evaluated. For this, a urine flow rate of 1 mL min^{-1} and pH 7.0 was applied as predefined conditions. From Fig. 4.3, panel B, it can be noted that using NC membrane with pore size equal to 0.22 μm, the extraction significantly increased, protein recovery being equal to 79 ± 4% (n = 3), while, using NC membrane with pore size equal to 0.45 μm, the recovery was approximately 70 ± 6% (n = 3). This observation can be explained by the fact that the smaller pores offer good accessibility and large areas of surfaces, allowing a greater adsorption of proteins in the membrane [6]. In other words, the pores in the membranes can be referred to as charged capillaries. The different pore size eventually contributes to the available amount of such charged capillaries in a membrane sample, so the smaller pores (with their higher number of charged capillaries) are able to bind more protein molecules to the membrane because of their high amount of active site [5]. Therefore, these results show that the membrane pore has a significant effect on the extraction of the protein and thus, for the following experiment, a NC membrane with a pore size equal to 0.22 μm was selected.

4.4.1.3   Influence of the Flow of Urine Passing Through the Membrane

The urine flow through of the membrane may play an important role in membrane adsorption [17]. In this sense, the urine flow rate was also investigated. For this, a series of experiments were carried out, using the urine flow rate: 0.25, 0.5 and 1.0 mL min^{-1}. Briefly, 1 mL of urine was passed through the NC membrane (0.22 μm) at different times (1, 2 and 4 min). The urine pH was adjusted to 7.0. The results of these experiments are illustrated in Fig. 4.3, panel C. As shown in the figure, the extraction efficiency increased significantly as the urine flow decreased. Thus, urine flow of 0.25 mL min^{-1} showed significant efficiency in extracting urinary protein when compared to 0.5 and 1 mL min^{-1}. For a urine flow rate of 0.25 ml min^{-1}, a protein recovery of $79 \pm 2\%$ (n = 3) was observed, while for 0.5 and 1.0 ml min^{-1}, the recovery were $71 \pm 3\%$ and 64 ± 3 (n = 3), respectively. This fact can be explained in terms of the interaction time between protein and membrane. In other words, when the urine flow is increased, the protein-membrane interaction decreases correspondingly, since the interaction time is small. However, when there is a decrease in the urine flow, the membrane-protein interaction time is increased, and consequently there is a great interaction between the urine proteins and the membrane surface, thus justifying the recovery of the urine flow of 0.25 mL min^{-1} [4]. Thus, 0.25 mL min^{-1} was selected as optimal urine flow for the following experiments.

4.4.1.4 Influence of pH

The pH value is an important parameter that can affect the urine protein extraction by the NC membrane, since it influences the loading states of the proteins as well as change on the membrane surface charge, influencing therefore the interaction between the proteins and the NC membrane [15]. To assess the effect of pH on protein extraction, urine pH was adjusted, ranging from 3 to 7. For this, urine flow and pore size were set as 0.25 mL min^{-1} and 0.22 μm, respectively. Under these conditions, the recoveries were $93 \pm 4\%$, $89 \pm 2\%$, $82 \pm 8\%$, $83 \pm 8\%$ and $78 \pm 4\%$ (n = 3) for the following pH values: 3, 4, 5, 6 and 7.0, respectively (Fig. 4.3, panel d). In view of the results obtained, pH 3.0 was selected as the optimal pH for extraction of urinary proteins by NC membrane.

4.4.2 Application: Gender Classification

Since the optimal conditions for the recovery of urine proteins using NC polymer membrane were found and the retention of the urine protein content was higher than 90%, the best condition of urine protein extraction was applied to obtain the profile electrophoresis of proteins extracted from the membrane (Fig. 4.4). With the analysis of SDS-PAGE, it is possible to complement the quantification analyzes, as well as to confirm the efficiency of the methodology developed here. In addition, a proof of concept was performed: a gender classification.

For the gender classification, the urine of male and female volunteers, as described in Sect. 4.3.2.1, was submitted to the sample preparation protocol developed in this thesis. The protein content of each urine sample was retained in the membrane, and then trypsin was applied, as explained in Sect. 4.3.2.4. Then the extracts containing the peptides were removed and prepared for MALDI analysis. Figure 4.5 shows some specimens of male (panel a) and female (panel b) spectra used for gender classification. From the figure, it can observe differences in the intensities of the spectra. In the male spectrum, mass/charge ratio (m/z) peaks 982, 1129, 1382, 1797 and 1910 are shown with high intensities when compared to the female spectrum, while for the

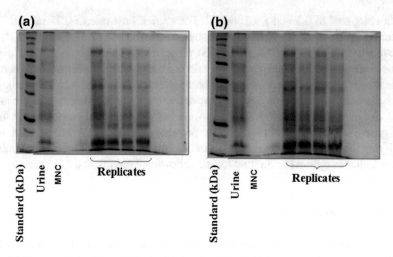

Fig. 4.4 Representative image of SDS-PAGE gel, obtained after desorption of the proteins from the NC membrane surface. **a** first replica of the gel; **b** second gel replica

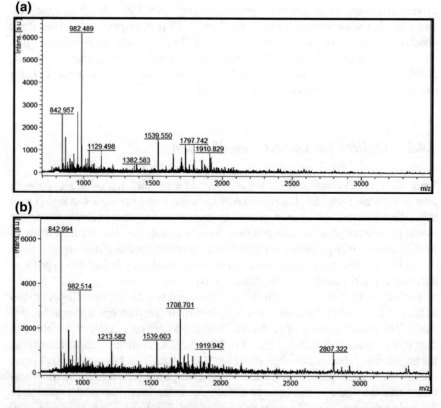

Fig. 4.5 Representative images of spectra obtained for gender classification. **a** male; **b** female

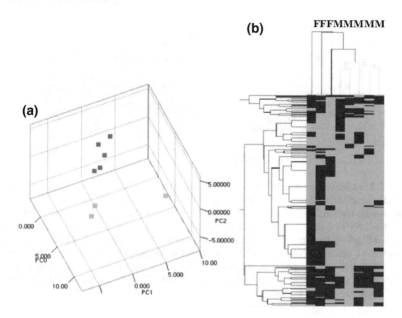

Fig. 4.6 Gender classification (male and female). **a** Principal component analysis (PCA), where red represents the male sex and green represents the female sex; and **b** hierarchical clustering, where F—female and M—male

female spectrum, the mass/charge ratio (m/z) peaks 842, 982, 1213, 1708, 1919 and 2807 are shown at higher intensities than the male spectrum, demonstrating therefore differences in the proteome between the sexes. Such differences may be justified in terms of biological variability. In other words, women have biological cycles during the month that increases the levels of hormones in their body, which can trigger differences in the proteome of their body fluids, especially urine.

The lists of obtained m/z were used to classify the samples using the PCA analysis. PCA analysis is shown in Fig. 4.6, panel a, showing a good classification, with males clearly grouped in addition to females. This was also confirmed using clustering as shown in Fig. 4.6, panel b, where clearly males and females are grouped separately.

To confirm the differences observed in the profiles obtained by MALDI-MS, an LC-MS/MS assessment was performed. From this evaluation, a total of 152 proteins and 159 proteins were identified for urine samples from the female group and the male group, respectively (Table A.1, Appendix). The Vulcan plot analysis depicted in Fig. 4.7, Panel a, shows 13 proteins with a significantly different concentration between female and male sex ($p < 0.05$). Three proteins were found with higher concentration in the female sample while ten proteins were found to be positively regulated in the male (Fig. 4.7, panel B, Table A1, Appendix). This fact reflects gender differences and helps explain why gender can be discriminated using urine and the MALDI-based mass spectrometry profile. The remaining 149 proteins had similar concentrations between female and male urine samples ($p < 0.05$) (Table A.1, Appendix).

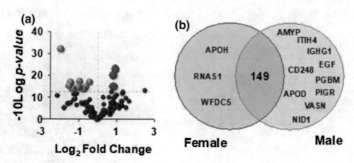

Fig. 4.7 Comparison of proteins obtained from female and male urine (pools): **a** Volcano plot illustrates abundantly differentiated proteins ($p < 0.05$). **b** Considering the common proteins: three proteins with high significant abundance in female ($p < 0.05$) and 10 proteins with high significant abundance in males ($p < 0.05$). Center: number of proteins with similar concentration levels. Where, APOH—Beta 2-glycoprotein1; RNAS1—pancreatic ribonuclease; WFDC5—Wap four-disulfide core domain protein 5; AMYP—Pancreatic alpha-amylase; ITIH4—Inter-alpha trypsin inhibitor heavy chain H4; ApoD—Apolipoprotein D; IGHG1—Ig gamma-1 chain C region; EGF—Pro-epidermal growth factor; CD248—Endosialin; PIGR—Polymeric immunoglobulin receptor; NID1—Nidogen-1; VASN—Vasorin

4.5 Partial Conclusion

In this thesis chapter was presented a new methodology that allows analyzing the proteome of urine. The method is based on the extraction of proteins from the urine through its interactions with cellulose-based polymer membranes. For this, some parameters were optimized: composition of the polymer membrane (cellulose, cellulose acetate, nitrocellulose), membrane pore size (0.22 and 0.45 μm), urine flow rate (0.25, 0.5 and 1 mL^{-1}) and pH of the medium (3–7). It was observed that the best results were obtained using nitrocellulose membrane, with a pore size of 0.22 μm and urine flow rate of 0.25 ml min^{-1}, in addition to using a pH 3 to obtain retention more than 90% of the urine proteins in the membrane. In addition, the methodology developed was applied in a gender classification study in order to evaluate a proteomic differentiation between males and females. From this study a significant classification was obtained between the sexes, according to the data of PCA and hierarchical clustering, as well as Volcano plot and the graphs of intersection.

References

1. Barratt J, Topham P (2007) Urine proteomics: the present and future of measuring urinary protein components in disease. Can Med Assoc J 177(4):361–368
2. Decramer S, Gonzalez de Peredo A, Breuli B, Mischak H, Monsarrat B, Bascands JL, Schanstra JP (2008) Urine in clinical proteomics. Mol Cell Proteom 7(10):1850–1862
3. de Jesus JR, Santos HM, López-Fernández H, Lodeiro C, Arruda MAZ, Capelo JL (2018) Ultrasonic-based membrane aided sample preparation of urine proteomes. Talanta 178:864–869

4. Low SC, Shaimi R, Thandaithabany Y, Lim JK, Ahmad AL, Ismail A (2013) Electrophoretic interactions between nitrocellulose membranes and proteins: biointerface analysis and protein adhesion properties. Colloids Surf B 110:248–253

5. Na N, Liu T, Yang X, Sun B, Ouyang J, Ouyang J (2012) A simple cellulose acetate membrane-based small lanes technique for protein electrophoresis. Anal Bioanal Chem 404(3):753–762

6. Pimet F, Perez E, Belfort G (1995) Molecular interactions between proteins and synthetic membrane polymer films. Langmuir 11:1229–1235

7. Remer T, Montenegro-Bethancourt G, Shi L (2014) Long-term urine biobanking: storage stability of clinical chemical parameters under moderate freezing conditions without use of preservatives. Clin Biochem 47(18):307–311

8. Saad A, Hanbury DC, McNicholas TA, Boustead GB, Morgan S, Woosman AC (2002) A Study comparing various noninvasive methods of detecting blader cancer in urine. BJU Int J 89:369–373

9. Scher MS, Ludington-Hoe S, Kaffashi F, Johnson MW, Holditch-Davis D, Loparo KA (2009) Neurophysiologic assessment of brain maturation afer an 8-week trial of skin-to-skin contact on preterm infants. Clin Neurophysiol 120:1812–1818

10. Sigdel TK, Lau K, Schilling J, Sarwal M (2008) Optimizing protein recovery for urinary proteomics, a tool to monitor renal transplantation. Clin Transplant 22(5):617–623

11. Spahr CS, Davis MT, McGinley MD, Robinson JH, Bures EJ, Beierle J, Mort J, Courchesne PL, Chen K, Wahl RC, Yu W, Luethy R, Patterson SD (2001) Towardsdefining the urinary proteome using liquid chromatography-tandem mass spectrometry. I. Profiling an unfractionated tryptic digest. Proteomics 1:93–107

12. Thongboonkerd V (2007) Practical points in urinary proteomics. J Proteome Res 6(10):3881–3890

13. Thongboonkerd V, Chutipongtanate S, Kanlaya R (2006) Systematic evaluation of sample preparation methods for gel-based human urinary proteomics: quantity, quality, and variability. J Proteome Res 5:183–191

14. Thongboonkerd V, Malasit P (2005) Renal and urinary proteomics: current applications and challenges. Proteomics 5:1033–1042

15. Thongboonkerd V, Mungdee S, Chiangjong W (2009) Should urine pH be adjusted prior to gel-based proteome analysis? J Proteome Res 8:3206–3211

16. Yamamoto T, Langham RG, Ronco P, Knepper MA, Thongboonkerd V (2007) Towards standard protocols and guidelines for urine proteomics: a report on the human kidney and urine proteome project (HKUPP) symposium and workshop, Seoul, Korea, 6 Oct 2007 and 1 Nov 2007, San francisco, CA, USA. Proteomics 8(11):2156–2159

17. Zhang Z (2008) Influence of flow separation location on phonation onset. J Acoust Soc Am 124(3):1689–1694

Chapter 5
Final Conclusion

In this thesis, the profiles of proteins and metal ions in blood serum samples from healthy individuals and patients with bipolar disorder, schizophrenia and other psychiatric disorders, such as schizoaffective, were evaluated in order to identify possible candidate chemical species to biomarkers of bipolar disorder. In addition, this thesis has also presented a new methodology for sample preparation based on polymer membranes to characterize the interferon-free urine proteome.

In the first chapter, it was observed that among the depletion methods evaluated to simplify the blood serum proteome of the studied groups (healthy and diseased individuals), PM kit was presented as the best strategy to remove proteins of high abundance and therefore was chosen for the treatment of BD, SCZ, OD, HCNF and HCF serum. From the 2-D DIGE gel analysis, 37 protein spots were found to be differentially abundant ($p < 0.05$, Student's t-test). From these detected spots, 13 different proteins were identified: ApoA1, ApoE, ApoC3, ApoA4, Samp, SerpinA1, TTR, IgK, Alb, VTN, TR, C4A and C4B. Through the global interaction network, these proteins act together and are associated with proinflammatory cytokines, which act in response to some inflammatory lesion, such as neuroinflammation, justifying, therefore the alterations found in the blood serum sample of the patients with BD when compared to controls.

In the second chapter, a method based on ultrasound-assisted extraction was proposed to determine Zn, Cu, Fe, Li, Cd and Pb in human serum samples by ICP-MS to discriminate patients with bipolar disorder from other psychiatric disorders. The sonoreactor of the cup-horn type together with the extraction solution containing 40% (v/v) HNO_3 + 30% (v/v) HCl with short sonication times (3 min) and low sonication amplitude (60%, equivalent to power of 160 W) proved to be efficient for extracting the metals in serum samples. Following the application of the proposed method in serum samples from patients with BD, SCZ and healthy individuals, significant differences were observed in Fe, Cu, Zn and Li levels for BD and SCZ when compared to controls, and for group of BD, the metals were observed at high level, while for the group of SCZ, all metals were found at low levels.

© Springer Nature Switzerland AG 2019
J. R. de Jesus, *Proteomic and Ionomic Study for Identification of Biomarkers in Biological Fluid Samples of Patients with Psychiatric Disorders and Healthy Individuals*, Springer Theses, https://doi.org/10.1007/978-3-030-29473-1_5

After the two omic studies (proteomics and ionomics), a strong relationship has been established between the results obtained. For example, significant differences were observed in iron levels for both diseases (BD and SCZ) when compared to the control. Iron is essential for normal neurological function, effectively participating in the synthesis of myelin and neurotransmitters. The high level of iron concentration in the patients brain with neurological diseases is directly related to increased transferrin protein in some brain regions. This information corroborates with the results found in this thesis, since a high abundance of transferrin and free iron ions was observed in blood serum of patients with BD when compared with healthy individuals. Another important relation between the two studies (proteomic and ionomic) is related to the zinc and copper ions found significantly altered in this thesis study for BD and SCZ when compared to the control. Both metal ions (Zn and Cu) can interact with amyloid proteins. An increase in the level of zinc and serum amyloid P protein components was observed in a sample of patients with BD when compared to patients in the schizophrenia group, thus confirming the direct relationship between proteins and metals in the pathophysiology of bipolar disorder.

In the third chapter, it was observed that the polymer-membrane-based method of nitrocellulose to characterize the urine proteome in samples from healthy individuals allowed a recovery of 90% of the proteins present in the fluid, thus presenting as an efficient method of extraction of urine proteins. It was also observed that, with the application of this methodology developed in urine samples from healthy individuals to differentiate the proteome from the male and female sexes, a significant classification was established according to statistical data, being able to affirm that there are differences in the proteomes between the sexes.

As possible limitations of this study, we cite the sample number involved in the research. However, this fact is justified in terms of the selection of the participants, since homogenization was sought in the profiles of the individuals involved in the study, both age, sex and weight, as well as drugs applied to the treatment of the disease.

In view of the results obtained and the information presented, this thesis opens new perspectives for the understanding of the pathophysiological mechanisms of bipolar disorder, as well as presents a new sample preparation procedure in body fluids, specifically blood serum and urine, with the objective to discover biological markers of human diseases.

Appendix

CONSENT TERM

Research data

Title of the project: "Proteomic and metalomic study for the identification of possible biomarkers in serum samples from patients with bipolar affective disorder"
Researcher in charge: Alessandra Sussilini
Position / Function: Professor at the Chemistry Institute of Unicamp

I am part of the research project that aims to analyze and compare blood proteins and metalloproteins in people with bipolar disorder or not (control group), in order to verify if there is a relationship between these proteins and such, with treatment applied or with genetic factors.

In order to conduct this research, I need you to agree to a brief questionnaire regarding personal data, the use of medicines and the presence of diseases, as well as donating 5 ml of blood to be analyzed. In addition, we also invite you to participate in an interview, answering questions about your current mental health, health problems presented in the past by you or family and treatments performed. By agreeing to participate in the survey, qualified professionals will collect blood.

The absolute secrecy of your identity is guaranteed in the presentation of the results obtained during the research. If there is interest, I can give you the individual result of your blood analysis. There will be no financial benefit. The discomfort and possible risk of participation refer to blood collection, but this will be performed by qualified professionals, using suitable and disposable material. It is also guaranteed that you may withdraw from participating in the survey at any time, without prejudice to your current or future treatment, and that, if I do not authorize, the samples collected will not be used for other purposes (they will only be used in accordance with the objectives of this work).

© Springer Nature Switzerland AG 2019

J. R. de Jesus, *Proteomic and Ionomic Study for Identification of Biomarkers in Biological Fluid Samples of Patients with Psychiatric Disorders and Healthy Individuals*, Springer Theses, https://doi.org/10.1007/978-3-030-29473-1

You will receive a copy of this document and I will make myself available for your follow-up to the collection laboratory and also for necessary explanations before, during and after the research.

I declare that, after being clarified by the researcher and having understood what was explained to me, I agree to participate in the Research Project. I authorize the storage of my blood by IQ-Unicamp for use in future research, as long as they are duly approved by FCM-Unicamp's Research Ethics Committee (CEP) and, if applicable, by the National Council for Research Ethics (CONEP):

_____Yes   ____No

Campinas, _____ of _____ of 20___.

_____   _____

Donor's signature   Researcher's signature

DATA COLLECTION FORM

ALL GROUPS:

1) Research identification: Group/Number: _____

(to be completed by the researcher)

2) Age: _____ years

3) Height: _____

4) Weight: _____

5) Are you a smoker? ( ) Yes   ( ) No

If so, for how long?_____ How many cigarettes a day?_____

6) Do you have any illness? ( ) Yes   ( )No

If yes, what? _____

7) Do you use medications (do not consider those used in the treatment of bipolar affective disorder, if any)? ( ) Yes ( )No

If yes: Which one? _____

What is the dose? _____

How long? _____

GROUPS: Bipolar disorder, schizophrenic, other mental disorders:

Regarding bipolar affective disorder:

6) How long has the disease been diagnosed? _____

7) Have you ever been diagnosed as depressive? ( ) Yes ( ) No

8) Which drug (s) is (are) used in your treatment?_____

9) What is the dose? _____

10) How long have you had this treatment? _____

11) Have you done any different treatment before today? ( ) Yes ( )No

If yes, with which drug (s)? What's the dose?

_____

Other comments or comments:

_____

_____

_____

See Table A.1.

Table A.1 Proteins identified for the female and male sex ($p < 0.05$)

Protein	Abundant factor (sex male:sexo female)
CD27 antigen	+1.5
Ceruloplasmin	−1.2
Protein shisa-5	+1.2
Ig kappa chain V-III region B6	−1.5
SLAM family member 5	+1.8
Autophagy-related protein 9B	−1.7
Protein AMBP	+1.2
Pepsin A	−1.5
Pancreatic secretory trypsin inhibitor	−1.4
Alpha-1-antitrypsin	+1.5
Alpha-1B-glycoprotein	−1.4
WNT1-inducible-signaling pathway protein 2	+1.5
Attractin	−1.5
Butyrophilin subfamily 2 member A1	+1.5
Ig kappa chain V-III region HAH	−1.3
Src substrate cortactin	+1.1
Collagen alpha-1(VI) chain	−1.6
Ig kappa chain V-III region WOL	+1.4
Filamin-C	−1.5
PH-interacting protein	+1.8
Small proline-rich protein 2A	−1.3
Protocadherin gamma-C3	+1.2
Leucine-rich alpha-2-glycoprotein	+1.5
Beta-galactosidase	+1.5
Ig kappa chain V-IV region Len	+1.3
Ig lambda chain C regions	−1.1
DNA-binding protein A	+1.5
Serine/threonine-protein kinase PAK 3	+1.8
Ig gamma-2 chain C region	−1.5
Phosphatidylethanolamine-binding protein 1	+1.2

(continued)

Table A.1 (continued)

Protein	Abundant factor (sex male:sexo female)
Protein delta homolog 1	+1.5
Lysosomal protective protein	−1.4
Vasorin	+2.5
Lysosomal alpha-glucosidase	+1.3
Tumor necrosis factor receptor superfamily member 1B	−1.1
Endonuclease domain-containing 1 protein	−1.5
Kallikrein-1	−1.2
Beta-2-microglobulin	+1.5
Fibrinogen alpha chain	−1.2
Uncharacterized protein KIAA0467	−1.5
Protein S100-A7	+1.4
Protein S100-A8	+1.5
Protein S100-A9	+1.7
Ig kappa chain V-II region Cum	−1.8
Folate receptor alpha	−1.3
Cadherin-13	+1.3
Tyrosine-protein kinase receptor UFO	−1.5
Alpha-2-HS-glycoprotein	−1.2
WAP four-disulfide core domain protein 5	−2.2
Low affinity immunoglobulin gamma Fc region receptor III-A	−1.5
Rho GTPase-activating protein 10	+1.6
Forkhead-associated domain-containing protein 1	+1.2
Cochlin	−1.2
Ig kappa chain V-III region SIE	−1.5
Dermatopontin	−1.4
Plasma serine protease inhibitor	−1.5
Signal-regulatory protein beta-1 isoform 3	−1.1
Cdc42 effector protein 4	+1.3
Deoxyribonuclease-1	−1.5
Thrombomodulin	−1.8
Insulin-like growth factor-binding protein 7	+1.5
Proapoptotic caspase adapter protein	−1.2
Ly6/PLAUR domain-containing protein 3	−1.5
Twisted gastrulation protein homolog 1	+1.4

(continued)

Table A.1 (continued)

Protein	Abundant factor (sex male:sexo female)
Lymphatic vessel endothelial hyaluronic acid receptor 1	−1.5
Desmocollin-2	+1.7
Nidogen-1	+2.8
Complement decay-accelerating factor	+1.3
Vesicular integral-membrane protein VIP36	+1.3
Mucin-like protein 1	−1.5
Polymeric immunoglobulin receptor	+2.6
Immunoglobulin J chain	−1.5
ICOS ligand	+1.5
Uncharacterized protein C1orf56	+1.7
Prothrombin	−1.6
Hemoglobin subunit alpha	−1.4
Mannosyl-oligosaccharide 1	+1.5
Tetranectin	−1.3
Endothelin B receptor	+1.5
Apolipoprotein A-II	−1.6
Pro-epidermal growth factor	+2.4
Vitronectin	−1.8
Elafin	+1.2
Fibronectin	+1.5
Clusterin	+1.4
Alpha-1-acid glycoprotein 1	+1.5
Ig kappa chain V-III region VH (Fragment)	+1.2
Keratin	+1.1
Alpha-1-acid glycoprotein 2	+1.5
Vitamin D-binding protein	−1.8
Prolactin-inducible protein	+1.3
Ig mu chain C region	−1.2
CD44 antigen	+1.5
Uteroglobin	+1.4
Cystatin-C	+1.5
GTPase IMAP family member 2	−1.4
Lactotransferrin	+1.8
Tumor necrosis factor receptor superfamily member 12A	−1.4
Podocalyxin-like protein 1	+1.5

(continued)

Table A.1 (continued)

Protein	Abundant factor (sex male:sexo female)
Cystatin-A	+1.3
Tyrosine-protein kinase ABL2	−1.1
Bone marrow proteoglycan	+1.5
Cystatin-M	+1.4
Granulins	−1.4
Metallothionein-1E	+1.2
Ig kappa chain V-I region Walker	+1.5
Secreted Ly-6/uPAR-related protein 2	−1.4
Secreted Ly-6/uPAR-related protein 1	+1.5
Mucosal addressin cell adhesion molecule 1	+1.3
Tubulointerstitial nephritis antigen-like	−1.1
Protein NOV homolog	−1.5
Glyceraldehyde-3-phosphate dehydrogenase	−1.8
Ig lambda chain V-I region VOR	+1.5
Lithostathine-1-alpha	−1.2
Fibulin-5	−1.5
Inter-alpha-trypsin inhibitor heavy chain H4	+2.4
Fibulin-2	+1.5
Putative lipocalin 1-like protein 1	+1.6
Ubiquitin	−1.8
Calmodulin-like protein 5	−1.3
Fibulin-1	+1.3
Ig gamma-1 chain C region	+2.5
Prostate-specific antigen	−1.3
Procollagen C-endopeptidase enhancer 1	+1.2
Poliovirus receptor	−1.5
Uromodulin	+1.6
Cadherin-2	+1.4
Thioredoxin	−1.2
Cadherin-1	−1.5
Ig lambda chain V-III region LOI	−1.4
Collagen alpha-1(XII) chain	−1.5
Ig lambda chain V-III region SH	−1.1
Cathelicidin antimicrobial peptide	+1.3
Actin	−1.5
Endosialin	+2.5
Ig gamma-4 chain C region	+1.3

(continued)

Table A.1 (continued)

Protein	Abundant factor (sex male:sexo female)
Tumor necrosis factor receptor superfamily member 10C	−1.2
SH3 domain-binding glutamic acid-rich-like protein 3	−1.5
Neuronal growth regulator 1	+1.4
Alpha-amylase 2B	−1.5
Cubilin	+1.6
Retinol-binding protein 4	+1.8
Connective tissue growth factor	+1.3
Cornifin-B	+1.3
Cornifin-A	−1.5
Protein S100-A7A	+1.6
Collagen alpha-1(XV) chain	−1.5
Histone demethylase UTY	+1.6
Serotransferrin	+1.5
Fatty acid-binding protein	+1.3
Arylsulfatase A	−1.1
Trinucleotide repeat-containing gene 6B protein	+1.5
Acyl-CoA-binding protein	+2.0
Matrilin-4	−1.4
Interleukin-6 receptor subunit beta	+1.2
Non-secretory ribonuclease	+1.5
Zinc finger protein 649	−1.4
Filaggrin	+1.5
Low-density lipoprotein receptor-related protein 2	+1.3
Leukocyte-associated immunoglobulin-like receptor 1	−1.1
Hemicentin-1	−1.5
High mobility group protein 1-like 10	−1.3
Ribonuclease pancreatic	−2.4
Complement C4-A	−1.2
Ig kappa chain V-I region AG	−1.5
Membrane protein FAM174A	+1.4
Monocyte differentiation antigen CD14	+1.5
Bile salt-activated lipase	+1.7
Fibrillin-1	−1.8
Extracellular superoxide dismutase [Cu-Zn]	−1.3
Complement C1r subcomponent-like protein	+1.3

(continued)

Table A.1 (continued)

Protein	Abundant factor (sex male:sexo female)
Ig kappa chain V-I region DEE	−1.5
Protein S100-A11	−1.2
Sodium- and chloride-dependent glycine transporter 2	+1.2
Xylosyltransferase 1	−1.5
Roundabout homolog 4	+1.6
Uncharacterized protein C5orf42	+1.4
Sterol regulatory element-binding protein 2	−1.2
Alpha-2-macroglobulin	−1.5
Plasminogen	−1.4
Myosin-8	−1.5
Platelet glycoprotein VI	−1.1
Signal-regulatory protein beta-1	+1.3
Ig kappa chain V-III region NG9 (Fragment)	−1.5
Putative tenascin-XA	−1.1
Structure-specific endonuclease subunit SLX4	+1.5
Vitelline membrane outer layer protein 1 homolog	−1.2
Latent-transforming growth factor beta-binding protein 4	−1.5
Glutaminyl-peptide cyclotransferase	+1.4
Superoxide dismutase [Cu-Zn]	−1.5
Cornulin	+1.3
Tyrosine-protein kinase Blk	+1.8
Golgi membrane protein 1	+1.3
Annexin A1	+1.3
Ig kappa chain C region	−1.5
Neurosecretory protein VGF	+1.6
Ig alpha-2 chain C region	−1.5
Myeloperoxidase	+1.5
Secreted and transmembrane protein 1	+1.2
Ig kappa chain V-I region Ni	+1.3
Gelsolin	−1.1
Major prion protein	+1.5
Ig heavy chain V-III region TIL	+1.3
Beta-2-glycoprotein 1	−2.8
Sorting nexin-30	+1.2
Alpha-1-antichymotrypsin	+1.5

(continued)

Table A.1 (continued)

Protein	Abundant factor (sex male:sexo female)
Plasma protease C1 inhibitor	−1.4
Macrophage colony-stimulating factor 1	+1.5
Complement component C7	+1.3
Complement component C6	−1.1
Probable serine carboxypeptidase CPVL	−1.5
Trefoil factor 2	−1.4
T-cell antigen CD7	+1.3
Glutamyl aminopeptidase	−1.2
Beta-defensin 1	−1.5
Phosphoinositide-3-kinase-interacting protein 1	+1.4
EGF-containing fibulin-like extracellular matrix protein 2	+1.5
DnaJ homolog subfamily B member 4	+1.5
Latent-transforming growth factor beta-binding protein 2	−1.8
EGF-containing fibulin-like extracellular matrix protein 1	−1.3
Ig lambda chain V-IV region Hil	+1.3
Biotinidase	−1.5
Hemopexin	−1.6
Aminopeptidase N	+1.7
Thrombospondin-1	−1.6
Beta-hexosaminidase subunit alpha	−1.4
Small proline-rich protein 3	+1.5
Haptoglobin	−1.3
Protein YIPF3	+1.5
Galectin-3-binding protein	−1.6
Peptidoglycan recognition protein 1	−1.5
Basement membrane-specific heparan sulfate proteoglycan core protein	−1.8
Activin receptor type-1B	+1.2
Hemoglobin subunit beta	+1.5
Mannan-binding lectin serine protease 2	+1.4
Ig gamma-3 chain C region	+1.5

(continued)

Table A.1 (continued)

Protein	Abundant factor (sex male:sexo female)
Serine protease inhibitor Kazal-type 5	+1.2
Osteopontin	+1.1
Prostaglandin-H2 D-isomerase	+1.5
ProSAAS	−1.8
Cell adhesion molecule 1	+1.3
Cell adhesion molecule 4	−1.2
Hepcidin	+1.5
Semenogelin-2	+1.4
Semenogelin-1	+1.5
Protein S100-A7-like 2	−1.3
Tumor susceptibility gene 101 protein	+1.8
Ig kappa chain V-I region HK102 (Fragment)	−1.4
Apolipoprotein(a)	+1.2
Cyclic AMP-responsive element-binding protein 3-like protein 3	−1.5
Phosphatidylcholine-sterol acyltransferase	+1.6
Sodium/potassium-transporting ATPase subunit gamma	+1.3
Zinc finger CCHC domain-containing protein 16	−1.2
Immunoglobulin superfamily member 8	−1.5
Laminin subunit gamma-1	−1.4
Tumor necrosis factor receptor superfamily member 14	−1.5
Guanylin	−1.1
Tumor necrosis factor receptor superfamily member 16	+1.3
Neutrophil gelatinase-associated lipocalin	−1.5
Zinc-alpha-2-glycoprotein	−1.1
Liver-expressed antimicrobial peptide 2	+1.2
Basigin	−1.2
Apolipoprotein D	+2.5
Apolipoprotein E	+1.4
Cathepsin B	−1.5
Maltase-glucoamylase	+2.0

(continued)

Table A.1 (continued)

Protein	Abundant factor (sex male:sexo female)
Protein FAM198B	+1.8
Cathepsin D	+1.3
Alpha-amylase 1	+1.3
Protein FAM71E1	−1.5
CD320 antigen	+1.6
Contactin-associated protein 1	−1.5
CD59 glycoprotein	+1.6
Endothelial protein C receptor	+1.5
Amyloid beta A4 protein	+1.3
CMRF35-like molecule 9	−1.1
CMRF35-like molecule 8	+1.5
Pancreatic alpha-amylase	+2.5
Frizzled-4	−1.8
Ig heavy chain V-III region BRO	+1.2
Serum albumin	+1.5
Ig kappa chain V-III region VG (Fragment)	−1.4
Interleukin-18-binding protein	+1.5
Kininogen-1	+1.3
Extracellular sulfatase Sulf-2	−1.1
Programmed cell death 1 ligand 2	−1.5
Kunitz-type protease inhibitor 2	−1.4
Ig alpha-1 chain C region	+1.2
E3 ubiquitin-protein ligase NEDD4	−1.2
Tenascin-X	−1.5

Printed in the United States
By Bookmasters